No Grid S Projects:

MW01224913

Step-by-Step DIY Projects for Off-Grid Self-Sufficiency with Advanced Guides and Practical Solutions for Every Situation

Jake Barrett

TABLE OF CONTENT

Introduction

Why Prepare for Off-Grid Self-Sufficiency?

In a world of increasing uncertainty, from economic instability to the ever-growing threat of natural disasters, the idea of being entirely self-sufficient has never been more relevant. Whether it's a power grid failure, a financial collapse, or extreme weather events, the modern world is becoming more unpredictable. Many people are beginning to realize that relying solely on modern infrastructure is a vulnerability, not a strength.

This book is designed for individuals who seek to break free from that dependency. Whether you're new to off-grid living or an experienced prepper looking to fine-tune your skills, *No Grid Survival Projects Bible* is your comprehensive guide to achieving true independence. You'll find practical, hands-on projects to ensure you and your family are prepared for any eventuality, no matter how severe.

What Does It Mean to Live Off the Grid?

Living off the grid doesn't just mean surviving in the wilderness. It means taking control of your resources—your energy, water, food, and security—so you don't have to depend on external systems to live. Off-grid living can look different for everyone: for some, it's about generating renewable energy for a homestead; for others, it's preparing a long-term emergency food supply or having the tools to defend your family if needed.

This guide will take you step-by-step through the essential skills you need to build a resilient, self-sustaining lifestyle. Whether you're preparing for a long-term crisis, wanting to reduce your carbon footprint, or simply aiming to live more independently, this book has you covered.

How to Use This Book

This book is structured to guide you through different areas of off-grid survival in a logical, practical manner. Each chapter focuses on a key aspect of off-grid living, from sourcing water to generating energy, and from growing your food to building a fortified home defense system. Every project is broken down into manageable steps, with detailed explanations, illustrations, and tips to ensure success.

- **For beginners**, the projects provide foundational knowledge to get started on your journey toward self-sufficiency.
- **For more experienced preppers**, the book includes advanced techniques and in-depth technical details to challenge and expand your skills.

This is not just a survival guide. It's a roadmap to a new way of life—one where you have full control over your environment, your resources, and your future.

Real-Life Success Stories: How Off-Grid Living Has Changed Lives

Throughout the book, you'll find real-life stories from people who have successfully transitioned to off-grid living. These stories are meant to inspire and provide you with insights into the challenges and rewards of adopting a self-sufficient lifestyle. You'll learn from those who have already made the leap and are thriving without reliance on the grid.

What You'll Learn

- **How to generate your own energy**: From solar power to wind turbines, you'll learn how to create sustainable energy sources that can power your home.
- **How to source and purify water**: Water is life. This book will teach you various methods to collect, filter, and store water to ensure you always have a reliable supply.
- **How to grow and preserve your own food**: Learn to grow an abundant garden, raise livestock, and preserve your harvest for long-term storage.
- **How to secure your home**: Defend your property and your loved ones with practical security measures and strategies for home fortification.
- **How to maintain health and hygiene off-grid**: Learn how to stay clean, treat injuries, and maintain your well-being without access to modern medical facilities.

A Path to Resilience

By the end of this book, you will have a comprehensive understanding of how to thrive independently of modern systems. Whether you aim to be fully off-grid or simply want to be better prepared for emergencies, this book offers projects that will build your resilience and help you protect what matters most—your family, your home, and your peace of mind.

1. Water Resource Collection and Management

1.1 Introduction to Off-Grid Water Systems

Water is essential for life, and ensuring a reliable, clean source of water is one of the primary challenges of off-grid living. Whether you're facing an emergency, preparing for long-term self-sufficiency, or just wanting to reduce your reliance on municipal water supplies, learning how to harvest, filter, and store water is crucial.

In an off-grid scenario, you cannot depend on a centralized water source. This means you need to have systems in place to both collect and purify water from natural sources. This section will cover the various methods of obtaining water, including rainwater harvesting, well construction, and alternative sources like plant transpiration. Additionally, we'll dive into the essential steps for purifying and storing water to ensure it's safe to drink.

Key Topics Covered:

- Your daily water requirements and how much you should store for emergencies
- Emergency water sourcing in crisis situations
- Off-grid water systems: Overview and setup
- Step-by-step guide to digging and constructing wells
- Techniques for rainwater collection using gutters and storage systems
- Purifying and filtering water with DIY systems
- Long-term water storage solutions

Ensuring a stable water supply is the foundation of any off-grid system. Once you understand the options available and how to implement them, you'll have taken the first step toward greater independence. This chapter will help you not only to secure your water needs but also to plan for future water demands, offering you and your family peace of mind in any situation.

1.1.1 Your Daily Water Requirements and Emergency Storage

Understanding your water needs is the first and most critical step in setting up an off-grid water system. On average, an adult requires about **1 gallon (3.8 liters) of water per day** for drinking alone. This amount does not account for other essential uses, such as cooking, hygiene, and cleaning. In an off-grid scenario, your total water requirements can quickly add up.

Daily Water Requirements Breakdown:

- **Drinking:** 1 gallon per person per day.
- **Cooking and food preparation:** 0.5 to 1 gallon per person per day.
- **Hygiene (washing hands, face, etc.):** 0.5 gallons per person per day.
- **Cleaning (dishes, basic laundry, etc.):** 1 gallon per person per day.

For a family of four, this can amount to **3 to 5 gallons (11 to 19 liters)** of water per person, per day, including all essential uses.

Emergency Water Storage

In preparation for emergencies or times when water sources may be scarce, it's essential to store an adequate amount of water. The **Federal Emergency Management Agency (FEMA)** recommends storing **at least a 2-week supply of water** for each person in your household. This means you should aim to store a minimum of **14 gallons (53 liters)** per person to cover two weeks of basic needs.

Important tips for water storage:

- **Containers:** Use food-grade containers, such as large plastic barrels or sealed water tanks, to store water. Make sure they are cleaned and disinfected properly before use.
- **Location:** Store water in a cool, dark place, away from direct sunlight to prevent the growth of bacteria and algae.
- **Rotation:** Replace stored water every 6 months to ensure freshness, or use water treatment options like purification tablets to extend the shelf life.

Key Takeaways:

- Calculate your total water needs based on the number of people and activities in your household.
- Store a minimum of **1 gallon (3.8 liters) per person per day** for drinking and at least **3 gallons (11 liters)** per person per day for total use.
- Plan for long-term storage and periodic rotation of your water supply to ensure it's always fresh and drinkable.

By calculating your water requirements and preparing for emergencies, you ensure that your family has a stable and safe water supply, even in situations where access to fresh water is limited. This foundational step in off-grid living helps eliminate the stress of water scarcity and prepares you for a variety of challenges.

1.1.2 Emergency Water Sources: How to Find and Access Water in a Crisis

In a survival or off-grid situation, you may find yourself without immediate access to a pre-stored water supply. Knowing how to locate and access emergency water sources is a vital skill that can make the difference between life and death in a crisis. Whether you're dealing with a natural disaster, a grid failure, or long-term off-grid living, understanding how to find water in your environment is critical.

1. Natural Water Sources

The most reliable sources of emergency water are natural ones, but they must be accessed and treated correctly to ensure they are safe for consumption. Below are some of the most common natural water sources you can tap into during an emergency:

- **Streams, Rivers, and Lakes:** These are easily recognizable sources of water in many regions. However, untreated surface water often contains bacteria, parasites, and other contaminants, so filtering and purifying this water before drinking is essential.
- **Rainwater:** Collecting rainwater is one of the simplest and safest ways to obtain clean water. You can easily gather rainwater from roofs using gutter systems and direct it into barrels for storage. However, be cautious of potential contaminants from roofing materials, and always filter and purify the water before use.
- **Snow and Ice:** If you live in a cold climate, snow and ice can provide an emergency water source. Always melt snow and ice before drinking—never eat snow directly, as it can lower your body temperature, leading to hypothermia.
- **Dew Collection:** Dew can be collected from plants and grass in the early morning by using a clean cloth or tarp. While this won't yield large amounts of water, it can be a valuable resource in arid regions.

2. Human-Made Water Sources

In an emergency, there may also be several human-made water sources around you that can provide short-term relief:

- **Water Heaters and Pipes:** In many homes, water heaters store a significant amount of water, which can be accessed during a crisis. Be sure to turn off the main valve to prevent contamination from entering the system.
- **Swimming Pools (for non-drinking purposes):** While not suitable for drinking due to chemicals like chlorine, pool water can be used for hygiene or cleaning in an emergency. If absolutely necessary, pool water can be purified with proper filtration methods, but this should be a last resort.
- **Toilet Tanks (non-bowl water):** The water stored in the toilet tank (not the bowl) is generally clean and safe for non-drinking purposes. Again, purification is recommended before consumption.

3. Alternative Water Collection Methods

In more extreme survival situations, when traditional sources are scarce, these alternative methods can be used to collect water:

- **Solar Still:** This method uses the sun's energy to evaporate water from soil, plants, or other sources, which is then collected in a container as condensation. It's particularly useful in arid environments. Solar stills are simple to build with basic materials, and they provide distilled water, which is safe for drinking.
- **Plant Transpiration Bags:** By covering a branch or leafy plant with a clear plastic bag, you can collect water that the plant releases through transpiration. This method works well in sunny conditions and can yield small but life-saving amounts of water.

4. Key Considerations for Accessing Emergency Water

- **Purification Is Critical:** Regardless of the source, most water you collect in an emergency will need to be purified to ensure it's safe for consumption. Boiling, chemical treatment (using water purification tablets or bleach), or using a water filter like a LifeStraw or Berkey are essential steps to make water potable.

- **Avoid Polluted Sources:** Steer clear of water that has obvious signs of contamination, such as an unpleasant smell, discoloration, or proximity to industrial areas or sewage runoff. In extreme cases, it's better to go without than to risk ingesting dangerous toxins.

5. Quick Emergency Water Checklist:

- Always prioritize **natural sources** like rainwater, rivers, and streams.
- Use **human-made sources** only when natural water is unavailable, and always purify them.
- Consider building **simple water collection systems** like solar stills and plant transpiration setups in long-term survival scenarios.
- **Purify all collected water** before drinking.

By understanding where and how to find emergency water sources, you can greatly increase your chances of staying hydrated and healthy during a crisis. The techniques and knowledge covered here will ensure that even in the worst scenarios, you have access to one of the most essential resources for survival.

2. Water Collection and Storage Techniques

When living off the grid, securing a sustainable and reliable water supply is paramount. Off-grid water systems are designed to capture, store, and purify water without relying on municipal sources. These systems can be as simple as collecting rainwater or as complex as drilling a well with solar-powered pumps. Understanding the various options available and how to set them up properly will ensure that you and your family have access to clean, safe water at all times.

In this section, we'll explore different off-grid water systems, including rainwater harvesting, well construction, and greywater recycling. We will also discuss the equipment and techniques necessary for setting up these systems, along with their advantages and disadvantages.

2.1 Rainwater Harvesting Systems

Rainwater harvesting is one of the most practical and sustainable methods for collecting water in an off-grid setting. By capturing rainwater from roofs or other surfaces, you can create a reliable water supply for drinking, cooking, gardening, and cleaning. Rainwater is relatively pure compared to other natural sources, but it still requires proper filtration and storage to ensure it's safe for consumption.

How Rainwater Harvesting Works

Rainwater harvesting involves collecting rain as it falls onto your roof or other structures and channeling it into storage tanks or barrels. This system is simple to set up and can provide a significant portion of your household's water needs, depending on your location and rainfall patterns.

The basic components of a rainwater harvesting system include:

1. **Catchment Area**: Typically, the roof of your home or another large surface. The larger the area, the more rainwater you can collect.
2. **Gutters and Downspouts**: These direct rainwater from the roof into the storage system. Gutters should be properly cleaned and maintained to prevent blockages and contamination.
3. **First Flush Diverter**: This device removes the first few gallons of rainwater that wash off the roof, which may contain dirt, leaves, or bird droppings. Once the initial runoff is diverted, cleaner water can flow into the storage tank.
4. **Storage Tanks or Barrels**: Rainwater is stored in large containers such as barrels, cisterns, or tanks. These should be covered and shielded from sunlight to prevent algae growth and contamination.
5. **Filtration System**: Although rainwater is generally clean, it's important to filter and purify it before use, especially for drinking and cooking. Filtration systems can range from simple DIY filters to more advanced setups like Berkey or ceramic filters.

Step-by-Step Guide to Setting Up a Rainwater Harvesting System

Step 1: Evaluate Your Catchment Area

Measure the size of your roof or catchment area to estimate how much water you can collect. On average, **1 inch of rain on 1,000 square feet of roof can yield approximately 600 gallons of water**. Make sure your roof is made of non-toxic materials, as rainwater can pick up chemicals from certain types of roofing.

Step 2: Install Gutters and Downspouts
Ensure that gutters are installed along the edges of your roof to capture rainwater. Position downspouts so that they direct water into your storage system. Be sure to install mesh screens at the top of the downspouts to prevent leaves and debris from entering the system.

Step 3: Set Up a First Flush Diverter
The first flush diverter ensures that the initial rain, which is likely to be contaminated, doesn't enter your storage system. Install the diverter on the downspout, so it automatically channels the first few gallons of water away from the tank. Once the diverter fills, cleaner water flows into your storage.

Step 4: Choose Your Storage Solution
Select tanks or barrels based on the amount of water you plan to collect and store. Storage options can range from **50-gallon rain barrels** to **larger cisterns that hold thousands of gallons**. Be sure to cover the storage containers to keep out debris, insects, and sunlight.

Step 5: Install a Filtration System
Water collected from rain can still contain contaminants from the air or roof. Set up a filtration system to clean the water before it's used for drinking, cooking, or bathing. Options include simple mesh filters, ceramic filters, or DIY systems like a sand filter. You may also consider adding a UV purifier or boiling water before consumption.

Step 6: Plan for Overflow
During heavy rains, your storage system may fill up quickly. Make sure you have an overflow system in place to direct excess water away from your foundation. This can be as simple as an additional hose leading to a garden or drainage area.

Benefits of Rainwater Harvesting

- **Cost-Effective:** Once installed, a rainwater harvesting system can significantly reduce your water bills or eliminate them entirely if you're living off-grid.
- **Sustainable:** It's an environmentally friendly way to collect water that would otherwise go unused, reducing your reliance on municipal water supplies.
- **Low Maintenance:** Aside from regular cleaning of gutters and the storage system, rainwater harvesting requires minimal upkeep.
- **Flexible Use:** Rainwater can be used for a variety of purposes, from watering plants to washing clothes, and with proper filtration, it can also be used for drinking.

Challenges and Considerations

- **Climate Dependent:** Rainwater harvesting is most effective in regions that receive regular rainfall. In areas prone to drought or dry seasons, additional water sources may be necessary.
- **Initial Cost:** Setting up a system with proper tanks, gutters, and filtration may require an upfront investment, though it quickly pays off in the long term.

- **Contamination Risks:** While rainwater is generally clean, it can become contaminated by roofing materials, air pollution, or animal waste. Proper filtration and maintenance are essential to ensure safe water for drinking and household use.

Key Takeaways:

- A **1-inch rainfall on a 1,000 square foot roof** can collect about **600 gallons** of water, making rainwater harvesting an efficient method for off-grid living.
- Ensure your storage tanks are covered and your gutters are clean to prevent contamination.
- Always filter and purify rainwater before using it for drinking or cooking to ensure it's safe for consumption.
- In areas with unpredictable rainfall, consider using rainwater harvesting in combination with other water sources, such as wells or greywater systems.

By setting up a proper rainwater harvesting system, you'll take a significant step toward ensuring a sustainable and self-sufficient water supply for your off-grid lifestyle.

2.2 Well Construction: How to Build and Maintain Your Own Off-Grid Well

For those living off-grid or in rural areas, a well is one of the most reliable ways to secure a long-term, sustainable water supply. Wells tap into groundwater and provide a steady source of clean water, even in regions where rainwater harvesting or surface water collection may not be feasible. However, constructing a well requires careful planning, proper tools, and an understanding of the water table in your area.

In this section, we will explore different types of wells, their construction methods, and the maintenance needed to ensure that your well remains a dependable water source for years to come.

Types of Wells

There are three primary types of wells used in off-grid water systems:

1. **Hand-Dug Wells**
2. **Drilled Wells**
3. **Driven Wells**

Each type has its own advantages and challenges, depending on the depth of the water table, the tools available, and your budget.

Hand-Dug Wells

A hand-dug well is the most traditional and labor-intensive method of well construction. These wells are typically shallow, reaching depths of **15 to 30 feet**. They are dug manually with shovels, picks, and other tools.

- **Advantages:** Hand-dug wells are relatively inexpensive and do not require heavy machinery. They are ideal for locations where the water table is shallow.
- **Disadvantages:** Due to their shallow depth, hand-dug wells are more prone to contamination from surface runoff. They may also run dry during droughts or seasonal changes.

Drilled Wells

Drilled wells are created using specialized drilling equipment and can reach depths of **100 to 500 feet or more**. These wells access deep aquifers, which provide cleaner and more reliable water compared to shallow wells.

- **Advantages:** Drilled wells reach deeper, cleaner groundwater, and are less susceptible to surface contamination. They provide a consistent water supply even during dry periods.
- **Disadvantages:** The cost of drilling a well can be significant, often requiring professional contractors and machinery. Maintenance and repairs may also require specialized equipment.

Driven Wells

Driven wells are created by driving a pipe with a pointed end (called a well point) into the ground until it reaches the water table. This method is generally used for shallow wells in sandy or soft soil.

- **Advantages:** Driven wells are relatively simple and inexpensive to install, making them an accessible option for DIY projects. They work well for shallow aquifers with soft soil.
- **Disadvantages:** Similar to hand-dug wells, driven wells are shallow and can be prone to contamination and running dry during droughts.

Step-by-Step Guide to Building a Driven Well

For many off-grid homesteaders, building a driven well is the most practical and affordable option. Here's how you can set one up on your property:

Step 1: Choose the Right Location

- Before you begin, it's essential to assess the groundwater availability in your area. Use local resources, such as geological surveys or consult a professional, to determine the water table depth.
- Avoid areas close to septic systems, livestock areas, or potential contaminants.

Step 2: Gather Your Tools and Materials

- **Well point**: A metal pipe with a pointed end and perforated sides to allow water to enter.
- **Drive cap**: A cap placed on top of the well point to protect it while driving it into the ground.
- **Drive pipe**: Sections of pipe that attach to the well point and extend the well deeper into the ground as needed.
- **Sledgehammer or driver**: Used to manually drive the well point into the ground.
- **Hand pump or solar-powered pump**: For drawing water from the well.

Step 3: Drive the Well Point into the Ground

- Attach the well point to the first section of drive pipe. Place the drive cap on top to protect the pipe.

- Use a sledgehammer or mechanical driver to drive the well point into the ground. Continue driving the pipe into the ground until you reach the water table. The depth will depend on the location, but most driven wells are between **15 to 30 feet** deep.

Step 4: Extend the Pipe as Needed

- As you drive the well point deeper, attach additional sections of drive pipe to extend the well. Use pipe couplings to connect each section securely.

Step 5: Test for Water

- Once you believe the well point has reached the water table, attach a hand pump or solar-powered pump to the top of the pipe. Begin pumping to check for water. If no water is found, you may need to drive the well deeper.

Step 6: Install the Pump

- After confirming water is present, install a permanent pump system. A hand pump is the most straightforward and requires no electricity, while a solar-powered pump offers a more automated solution for continuous water extraction.

Maintaining Your Well

Once your well is operational, regular maintenance is essential to ensure a steady and safe water supply.

- **Water Testing**: Periodically test your well water for contaminants such as bacteria, nitrates, and chemicals. This is especially important if your well is shallow or located near potential pollutants.
- **Pump Maintenance**: Regularly inspect and maintain your pump, whether it's hand-powered or solar-powered, to ensure it functions properly. Replace parts as needed.
- **Well Casing and Seal**: Ensure that the well casing remains intact and that the well is properly sealed at the surface to prevent contamination from entering the water supply.
- **Clean the Well**: Occasionally clean the well by removing any sediment or debris that may have entered. Use chlorine or other approved sanitizers to disinfect the well if needed.

Advantages of Building a Well for Off-Grid Living

- **Reliable Source of Water**: A well provides a consistent and reliable water supply, regardless of rainfall or surface water availability.
- **Cost-Effective Long-Term**: While the initial setup can be costly (especially for drilled wells), a well can save money in the long run by eliminating the need for municipal water or frequent water deliveries.
- **Self-Sustaining**: A well, especially when powered by a solar pump, offers complete water independence, making it ideal for off-grid living.

Disadvantages

- **Initial Cost**: Drilled wells can be expensive to install, especially if you need to hire professionals or use heavy machinery.
- **Maintenance**: Wells require ongoing maintenance, including pump repairs, water testing, and cleaning.
- **Contamination Risks**: Shallow wells are more vulnerable to contamination from surface water, chemicals, or nearby septic systems.

Key Takeaways:

- **Hand-dug wells** are low-cost but vulnerable to contamination and drying up.
- **Drilled wells** provide a deep, clean water source but require professional help and a higher upfront cost.
- **Driven wells** are an excellent DIY option for shallow water tables and offer a balance between affordability and accessibility.

By understanding the different types of wells and following the step-by-step guide, you can establish a reliable and independent water source for your off-grid property. Whether you opt for a simple driven well or a professionally drilled deep well, having access to groundwater is a cornerstone of off-grid living and self-sufficiency.

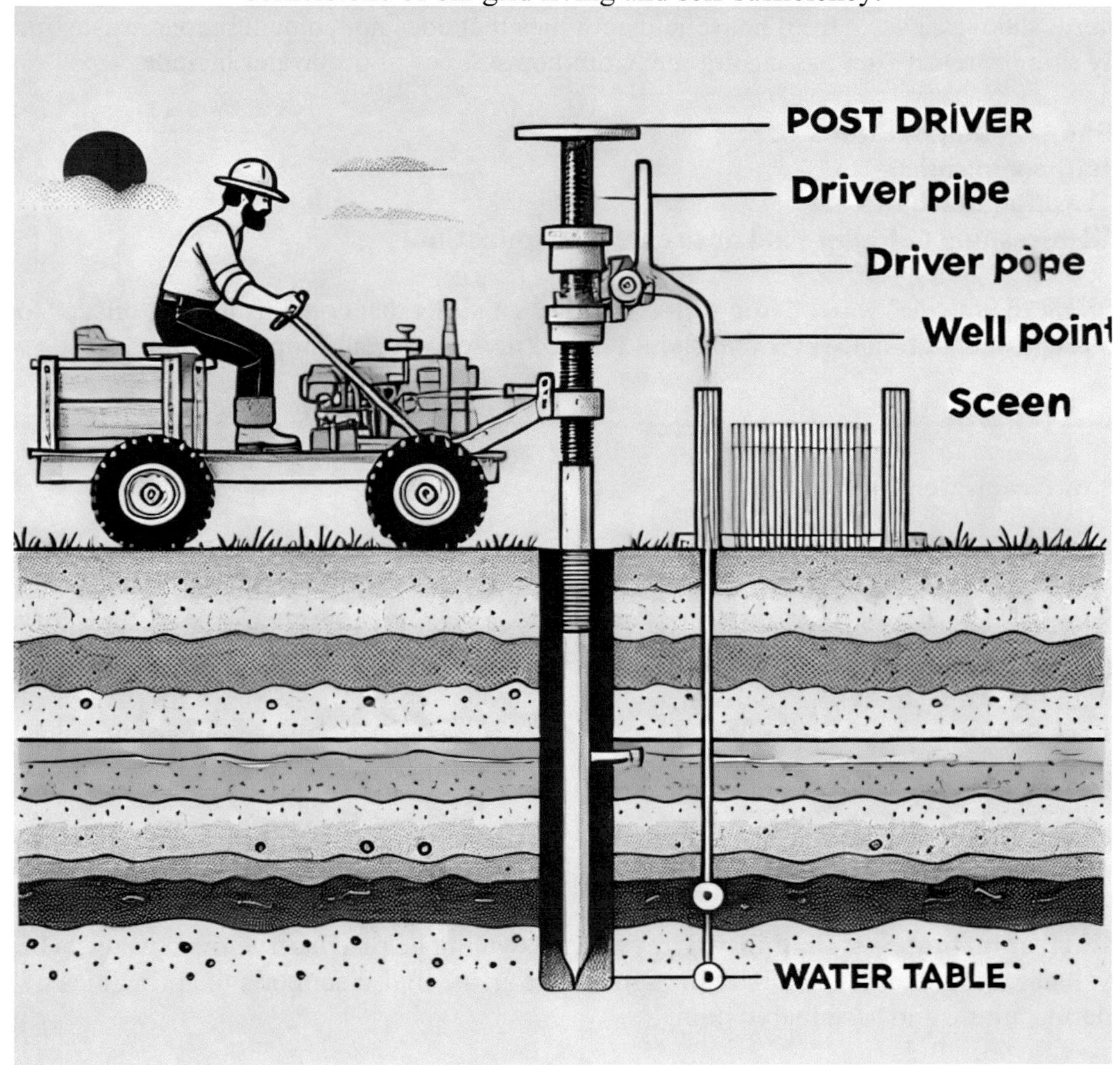

2.3 Greywater Recycling Systems: Reusing Water for Sustainable Off-Grid Living

Greywater recycling is an efficient and eco-friendly way to maximize your water use in an off-grid environment. Greywater refers to the relatively clean wastewater from household activities such as washing dishes, laundry, and showers. By diverting greywater for reuse in irrigation or other non-potable applications, you can significantly reduce your overall water consumption, making your off-grid water system more sustainable.

This section will explain how greywater systems work, the benefits of recycling greywater, and the steps to set up a simple DIY greywater system for your home or homestead.

What is Greywater?

Greywater is the wastewater from household activities that does not contain human waste, making it relatively safe for reuse after basic filtration. Common sources of greywater include:

- **Showers and bathtubs**
- **Bathroom sinks**
- **Washing machines**
- **Dishwashing (without food or grease contamination)**

It's important to note that **water from toilets or kitchen sinks** that contains grease, oils, or food particles is considered **blackwater** and should not be reused without proper treatment.

Benefits of Greywater Recycling

Greywater recycling offers several advantages for off-grid living, including:

- **Reduced Water Consumption:** By reusing greywater for irrigation and other non-potable purposes, you can lower your overall water usage, which is especially beneficial in regions with limited water resources.
- **Less Strain on Septic Systems:** Recycling greywater reduces the amount of wastewater that needs to be processed by your septic system, extending its lifespan and reducing maintenance needs.
- **Cost Savings:** By using greywater for tasks like watering plants or flushing toilets, you reduce the need to draw from your main water supply, leading to lower water bills or less dependency on other water sources.
- **Environmental Sustainability:** Greywater recycling helps conserve fresh water, which is a valuable resource, especially in drought-prone areas. It also supports the principles of permaculture and sustainable living.

How Greywater Systems Work

A greywater system diverts water from your household drains (such as from sinks or washing machines) and channels it for reuse. This water can then be used to irrigate gardens, flush toilets, or even clean outdoor surfaces.

The basic components of a greywater system include:

1. **Source:** The source is the household drain from which the greywater is collected (e.g., shower, washing machine).
2. **Diversion Device:** This device redirects greywater from your plumbing system to your greywater system instead of letting it flow to the septic tank or sewer.
3. **Filtration System:** Before the greywater is reused, it must be filtered to remove large particles like hair, lint, and soap residue. Filtration can be as simple as a mesh filter or more advanced systems with multiple stages of filtration.

4. **Distribution System:** The filtered greywater is then distributed to your garden or landscape via a drip irrigation system, or it can be stored temporarily for flushing toilets or other non-potable uses.

Step-by-Step Guide to Setting Up a Basic DIY Greywater System

A simple DIY greywater system can be set up using basic materials and is a cost-effective way to make your home more sustainable. Here's a basic guide to building a greywater system for garden irrigation:

Step 1: Identify Your Greywater Source

- Choose a reliable greywater source, such as your washing machine or bathroom sink. Washing machine greywater is one of the easiest and most practical sources for greywater reuse because the machine is typically already located near an exterior wall for easy diversion.

Step 2: Install a Diversion Device

- For a washing machine, install a greywater diverter valve on the discharge hose. This allows you to direct the water either into your greywater system or back into the sewer/septic system if needed. The valve should be easy to operate and accessible.

Step 3: Set Up Filtration

- Install a basic filter to remove larger debris. This can be a simple mesh filter placed in the diversion pipe or a more sophisticated multi-stage filter that removes finer particles. Filtration is essential to prevent clogging and contamination in your irrigation system.

Step 4: Build a Distribution System

- Once filtered, direct the greywater into a distribution system. For garden irrigation, the water should be channeled through pipes or hoses and released via drip irrigation or sub-surface watering methods to ensure that the water goes directly to the plant roots and minimizes contact with edible parts of plants.

Step 5: Irrigate the Right Plants

- Use greywater for ornamental plants, trees, and shrubs, or on non-edible plants in your garden. Avoid using greywater on vegetables or fruits unless it is delivered via subsurface irrigation to prevent contact with the edible parts of the plants.

Alternative Uses for Greywater

While irrigation is the most common use for greywater, there are other practical applications for off-grid living:

- **Flushing Toilets:** Greywater can be diverted into a storage tank for toilet flushing, reducing your reliance on fresh water for this purpose. This requires a pump or gravity-fed system to supply the water to the toilet cistern.
- **Non-Potable Cleaning:** Greywater can be used for cleaning outdoor surfaces like driveways, tools, or vehicles.
- **Laundry Pre-Rinse:** In some cases, greywater from a previous laundry cycle can be reused for the pre-rinse cycle of the next load, reducing overall water usage.

Safety and Regulations for Greywater Use

While greywater is relatively safe to reuse, it's important to follow some basic safety guidelines to avoid contamination or health risks:

- **Avoid Using Greywater on Edible Plants:** Unless it is delivered via a subsurface irrigation system, avoid applying greywater directly to plants that will be consumed to reduce the risk of bacteria or chemical exposure.
- **Use Biodegradable Products:** Ensure that the soaps, detergents, and cleaning products used in your home are biodegradable and eco-friendly, as harmful chemicals can damage soil and plants.
- **Check Local Regulations:** Some regions have specific regulations regarding greywater use. Be sure to check with local authorities to ensure that your greywater system complies with any legal requirements.

Key Considerations for Greywater Systems

- **Filtration:** Even though greywater is relatively clean, it still contains soap residues, oils, and small particles that need to be filtered before reuse. Ensure your system has an effective filtration mechanism in place.
- **Storage:** Greywater should not be stored for extended periods, as it can quickly become contaminated. Use it as soon as possible, ideally within 24 hours.
- **Maintenance:** Regularly check and clean your filters to prevent clogs and ensure the system operates efficiently. Inspect pipes and connections to avoid leaks or contamination.

Key Takeaways:

- Greywater recycling is a simple and effective way to reduce water consumption and live more sustainably in an off-grid environment.
- Setting up a basic greywater system for irrigation or non-potable uses can save fresh water and reduce strain on your primary water source.
- Always filter greywater before reuse, avoid applying it to edible plants, and ensure the system complies with local regulations.

By implementing a greywater recycling system, you can significantly reduce your off-grid water

3. Water Filtration, Purification, and Long-Term Storage

In an off-grid setting, it is crucial to ensure that the water you collect from natural sources like rainwater, wells, or greywater systems is safe for consumption. Without proper filtration and purification, water can harbor harmful contaminants such as bacteria, viruses, chemicals, and debris that pose serious health risks. Additionally, storing water correctly will help maintain its safety and quality for long-term use.

This section will guide you through the essential processes of filtering, purifying, and storing water in an off-grid environment. We will cover both DIY and commercial filtration options, purification methods, and the best practices for safe water storage.

3.1 Water Filtration: Removing Physical Contaminants for Safe Water Use

Water filtration is the first step in making water safe for use in an off-grid setting. Whether you are collecting rainwater, drawing water from a well, or sourcing it from a natural body of water, the filtration process is crucial for removing physical impurities like dirt, sand, leaves, and organic matter. By using proper filtration methods, you can ensure that the water is clean enough to be purified and stored for drinking, cooking, and other household needs.

Why Water Filtration is Important

Unfiltered water, even from seemingly clean sources, often contains various contaminants that can pose a risk to your health. Filtration helps to:

- **Remove debris** such as leaves, dirt, sand, and small stones.
- **Reduce sediment** that may cloud the water and clog purification systems.
- **Improve taste** by eliminating organic matter and particles that give water an unpleasant flavor or odor.

Proper filtration is essential for extending the lifespan of your water purification systems and ensuring that they work efficiently to remove smaller, more dangerous pathogens.

Types of Water Filtration Systems

There are several types of water filtration systems, each with its own benefits depending on the water source and your specific needs. The choice of filtration system can vary from simple DIY methods to advanced commercial systems.

1. Mesh Filters

Mesh filters are the simplest and most basic form of filtration. They use a fine mesh screen to block large particles like leaves, twigs, and dirt from entering your water storage system. These filters are often installed in rainwater harvesting systems or at the inlet of well pumps.

- **Advantages**: Inexpensive, easy to install, and suitable for pre-filtering large debris.
- **Disadvantages**: Does not filter smaller particles or pathogens; requires regular cleaning.

2. Sand Filters

Sand filtration is a highly effective, low-cost method that uses layers of sand and gravel to trap fine particles and some pathogens. This type of filter is commonly used in DIY setups and can be built using basic materials.

- **Advantages**: Inexpensive, effective at removing fine particles, and easy to build.
- **Disadvantages**: Requires regular maintenance to prevent clogging; does not filter out all pathogens.

3. Ceramic Filters

Ceramic water filters are widely used in both DIY and commercial systems. These filters have tiny pores that allow water to pass through but block bacteria, protozoa, and some viruses. Ceramic filters are often combined with activated carbon to remove chemical contaminants and improve the taste of water.

- **Advantages**: Highly effective at removing bacteria and sediment; reusable (can be cleaned and reused multiple times).
- **Disadvantages**: Slower filtration rate; may need regular scrubbing to maintain flow rate.

4. Carbon Filters

Activated carbon filters are highly effective at removing chemicals, chlorine, pesticides, and unpleasant odors from water. They work by absorbing organic compounds and contaminants, making the water taste and smell better.

- **Advantages**: Excellent for removing chemical contaminants and improving water taste; works well in combination with other filters.
- **Disadvantages**: Requires regular replacement or regeneration to maintain effectiveness; does not remove viruses or bacteria on its own.

5. 3-Bucket DIY Bio-Filter

This DIY method is simple, cost-effective, and highly practical for off-grid setups. The 3-bucket system uses multiple filtration layers (typically gravel, sand, and activated charcoal) to clean water progressively as it moves through each bucket. It is an excellent option for households that need to filter large volumes of water from natural sources.

How it Works:

- **Top Bucket**: The water is poured into the top bucket, which contains coarse gravel to trap large debris.

- **Middle Bucket**: The water then flows into the middle bucket, which contains fine sand to remove smaller particles and organic matter.
- **Bottom Bucket**: The final bucket contains activated charcoal or carbon to absorb chemicals, toxins, and impurities, providing clean, filtered water.

Advantages:

- DIY solution, easy to build using basic materials.
- Filters large volumes of water.
- Removes physical impurities, organic matter, and chemicals when properly maintained.

Disadvantages:

- Requires regular maintenance and replacement of the filter media.
- Does not remove bacteria or viruses unless combined with additional purification methods.

Choosing the Right Filtration System

When selecting a water filtration system, consider the following factors:

- **Water Source**: The type of water source (rainwater, well water, surface water) determines the level of filtration required. For example, rainwater collected from a roof will need debris removal, while well water may need sediment filtration.
- **Filtration Goals**: Decide whether you are primarily filtering for physical contaminants, chemical impurities, or biological pathogens. This will guide your choice of filtration method.
- **Flow Rate**: Depending on your household's water needs, you may require a fast or slow filtration system. Commercial filters may offer higher flow rates, while DIY systems might work more slowly but still meet daily needs.

Maintenance and Upkeep

For any filtration system to remain effective, regular maintenance is essential. This includes:

- **Cleaning filters**: Mesh and ceramic filters should be cleaned frequently to remove debris and prevent clogging. Sand filters should be backwashed or replaced periodically.
- **Replacing filter media**: Carbon filters and activated charcoal in DIY systems need to be replaced or regenerated regularly to maintain their ability to absorb chemicals.
- **Inspecting the system**: Regularly check your filtration setup for leaks, damage, or signs of wear that may reduce its effectiveness.

Key Takeaways:

- **Mesh and sand filters** are ideal for removing large debris and fine particles but should be paired with other systems to ensure comprehensive filtration.

- **Ceramic and carbon filters** provide higher levels of filtration, removing bacteria, chemicals, and other small contaminants.
- A **3-bucket DIY bio-filter** is a practical and cost-effective solution for filtering large volumes of water off-grid.
- Regular **maintenance** is essential to keep your filtration system working effectively and ensure safe water for your household.

By setting up an efficient filtration system and combining it with purification methods, you can guarantee that your off-grid water supply remains safe and clean for everyday use. Proper filtration is the foundation of a sustainable, reliable water system for any off-grid lifestyle.

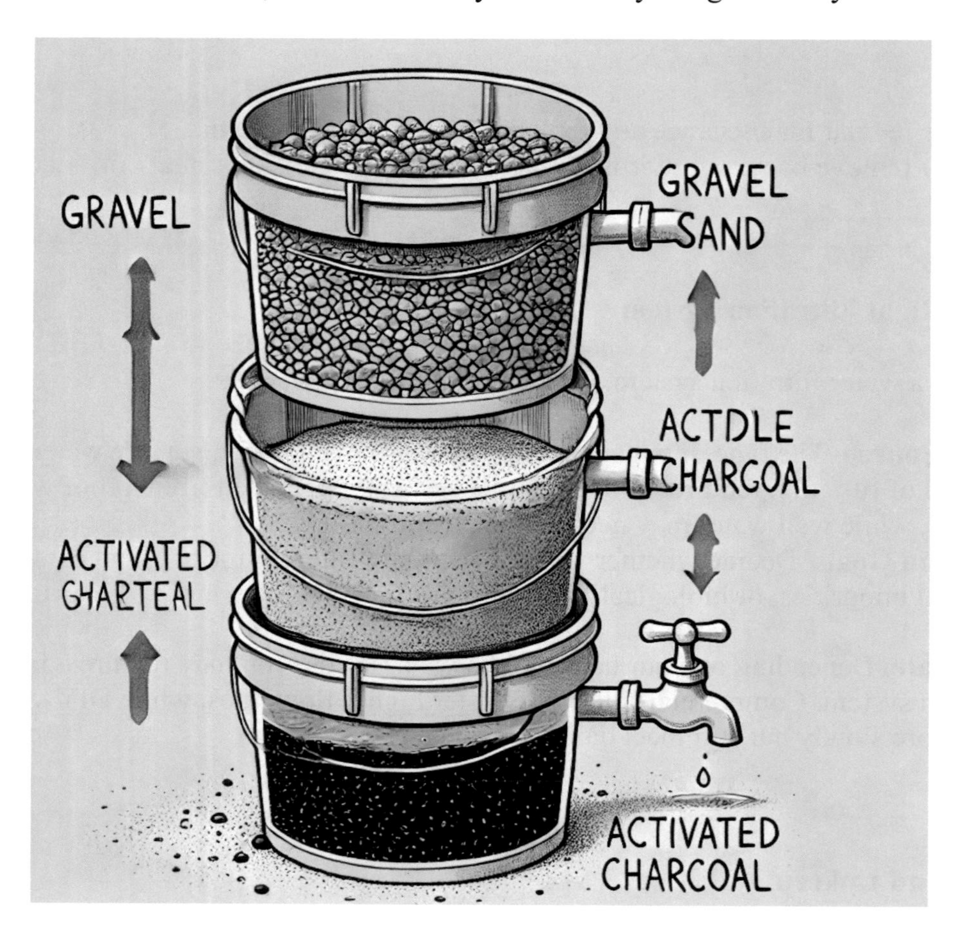

3.2 Water Purification: Making Water Safe for Drinking and Household Use

After filtering your water to remove physical impurities, the next critical step is purification. Filtration alone cannot remove all harmful pathogens like bacteria, viruses, and parasites. Water purification is essential to eliminate these microorganisms and ensure that your water is safe for drinking, cooking, and hygiene.

In this section, we will explore various water purification methods that can be used in an off-grid environment, from simple techniques like boiling to more advanced methods such as UV purification. Each method has its strengths and best-use scenarios, so understanding when and how to apply them is key to ensuring safe water for your household.

Common Water Purification Methods

1. Boiling

Boiling is one of the simplest and most effective methods for purifying water. By bringing water to a rolling boil for at least **5 to 10 minutes**, you can kill most harmful pathogens, including bacteria, viruses, and parasites.

- **How to Purify Water by Boiling:**
  - Bring water to a rolling boil for at least **5 minutes** (longer at higher altitudes).
  - Let the water cool before storing it in clean, covered containers.
- **Advantages:**
  - Highly effective at killing pathogens.
  - Requires no special equipment, just a heat source.
  - Ideal for emergency situations or when other methods are unavailable.
- **Disadvantages:**
  - Requires fuel or energy for boiling.
  - Does not remove chemical contaminants or improve water taste.
  - May not be practical for large volumes of water.

---

2. Chemical Treatment

Chemical water treatment involves using substances like **chlorine**, **iodine**, or **water purification tablets** to disinfect water. These chemicals kill pathogens by attacking their cell structures, rendering them harmless. Chemical treatments are particularly useful for treating water on the go or when access to boiling or filtration systems is limited.

- **Common Chemical Purification Methods:**
  - **Chlorine bleach**: Use **unscented household bleach** (containing 5-6% sodium hypochlorite). Add **8 drops per gallon** of water, stir, and let it sit for **30 minutes** before consuming.
  - **Iodine tablets**: Follow the manufacturer's instructions (typically one tablet per liter of water), and allow the water to sit for **30 minutes**.
  - **Purification tablets**: These are commercially available tablets containing chlorine dioxide or other disinfectants. They are convenient for treating small quantities of water quickly.
- **Advantages:**
  - Easy to use and portable.
  - Highly effective against bacteria and viruses.
  - Ideal for short-term or emergency use.
- **Disadvantages:**
  - May leave an unpleasant taste or odor in the water.
  - Does not remove physical contaminants or chemical pollutants.
  - Less effective against some parasites like Cryptosporidium.

---

3. Ultraviolet (UV) Purification

UV purification uses ultraviolet light to disinfect water by destroying the DNA of harmful microorganisms, preventing them from reproducing. UV water purifiers are available as handheld devices for small quantities of water or larger systems for household use. Some UV systems are powered by solar panels, making them suitable for off-grid living.

- **How UV Purification Works:**
  - Place the UV device into the water and activate it according to the manufacturer's instructions.
  - Stir the water as the UV light is emitted, ensuring that all the water is exposed to the light.
  - Most portable UV purifiers can treat **1 liter of water in 90 seconds**.
- **Advantages:**
  - Very effective against bacteria, viruses, and parasites.
  - Does not alter the taste or smell of water.
  - Solar-powered options available for off-grid use.
- **Disadvantages:**
  - Requires a power source (solar or batteries).
  - Does not remove physical contaminants, so pre-filtration is necessary.
  - Can only purify small amounts of water at a time with portable devices.

---

4. Solar Water Disinfection (SODIS)

Solar water disinfection, also known as **SODIS**, is an inexpensive and low-tech method for purifying water using the sun's UV rays. It involves placing water in clear plastic bottles and leaving them in direct sunlight for several hours. This method is particularly useful in areas with abundant sunlight and is often used in developing regions where access to clean water is limited.

- **How to Use SODIS:**
  - Fill clear plastic or glass bottles with water, leaving some air at the top to aid in oxygenation.
  - Place the bottles on a reflective surface (such as a metal sheet) in direct sunlight for at least **6 hours**.
  - After 6 hours, the water will be disinfected and safe to drink.
- **Advantages:**
  - Requires no fuel or chemicals.
  - Low-cost and easy to implement.
  - Ideal for emergency situations in sunny climates.
- **Disadvantages:**
  - Dependent on sunlight and weather conditions.
  - Only effective in small quantities (limited by bottle size).
  - Does not remove physical impurities or chemical contaminants.

---

5. Berkey Water Filtration System

The **Berkey Water Filtration System** is a popular and highly effective system for off-grid water purification. It combines filtration and purification into a single unit, removing bacteria, viruses, chemicals, heavy metals, and other contaminants. The Berkey system uses gravity to push water through ceramic filters that contain activated carbon, ensuring a thorough purification process.

- **How the Berkey System Works:**
  - Pour water into the top chamber, where it filters through the ceramic elements.
  - The filtered water collects in the bottom chamber, ready for use.
- **Advantages:**
  - Removes both physical and biological contaminants.
  - Long-lasting filters that can purify thousands of gallons of water.
  - Does not require electricity or batteries.
- **Disadvantages:**
  - Expensive compared to other methods.
  - Requires regular maintenance and cleaning of the filters.
  - Bulkier and less portable than other purification options.

Key Considerations for Water Purification

- **Source of Water**: Depending on your water source (rainwater, river, well, etc.), you may need a combination of filtration and purification methods. Surface water and rainwater often require more extensive purification than well water.
- **Type of Contaminants**: If your water is contaminated with chemicals or heavy metals, a method like **activated carbon** should be used in combination with other purification techniques to remove both biological and chemical contaminants.
- **Volume of Water**: Consider the amount of water you need to purify daily. Methods like **boiling** and **chemical treatment** are suitable for small quantities, while **UV systems** or the **Berkey filter** can handle larger volumes.

Key Takeaways:

- **Boiling** is a simple and reliable way to kill pathogens, but it does not remove chemicals.
- **Chemical treatments** like chlorine and iodine are portable and easy to use but may leave an unpleasant taste.
- **UV purification** is effective and preserves the taste of water but requires a power source.
- **SODIS** is a low-cost option that harnesses the power of sunlight, ideal for small amounts of water.
- The **Berkey system** offers comprehensive filtration and purification in one, but it requires a larger investment and regular maintenance.

By selecting the right purification method or combination of methods, you can ensure that your water is safe to drink and use, no matter where your off-grid lifestyle takes you.

3.3 Water Storage: Best Practices for Long-Term Water Storage in an Off-Grid Environment

Once you have filtered and purified your water, it's essential to store it properly to ensure that it remains clean and safe for future use. In an off-grid environment, having a reliable supply of water is crucial, especially during dry seasons, emergencies, or power outages that may prevent you from immediately accessing fresh water. Long-term water storage requires the right containers, storage conditions, and regular maintenance to ensure the water stays potable.

This section will guide you through the best practices for storing water, the types of containers to use, and how to ensure that your stored water remains safe over time.

Choosing the Right Containers for Water Storage

Selecting the right containers for water storage is a critical first step in ensuring the long-term safety and quality of your stored water. Not all containers are created equal, and using improper storage materials can lead to contamination, chemical leaching, or bacterial growth over time. When storing water for off-grid living, it's essential to choose food-grade, durable containers designed specifically for holding drinking water.

Food-Grade Containers

The best containers for long-term water storage are those made from materials approved for food and water storage. These materials are safe, non-toxic, and will not react with water over time. Here are the most common types of food-grade containers used for water storage:

1. Plastic Barrels and Tanks

- **55-Gallon Barrels**: One of the most popular options for large-scale water storage is the 55-gallon plastic barrel. These barrels are made from high-density polyethylene (HDPE), a durable, non-toxic plastic that is BPA-free and specifically designed for food and water storage. They are large enough to store a significant amount of water for long-term use but are compact enough to be stored in basements, garages, or outdoors with proper protection.
- **Tanks**: For households needing larger capacities, water storage tanks made of food-grade plastic can store hundreds or even thousands of gallons. These tanks are suitable for rainwater harvesting systems and can be installed above or below ground.
- **Advantages**: Lightweight, durable, affordable, and easy to clean. Plastic containers are widely available and come in a range of sizes, from small jugs to large tanks.
- **Disadvantages**: Plastic is susceptible to degradation over time if exposed to direct sunlight, which can weaken the container and potentially allow chemicals to leach into the water. This can be mitigated by storing plastic containers in a cool, dark location.

2. Glass Containers

- **Glass Jars and Bottles**: Glass is a highly reliable and inert material for storing water because it does not react with or leach chemicals into the water, making it one of the safest materials for long-term storage. Glass containers are typically used for smaller quantities of water, making them ideal for storing emergency drinking water or smaller batches of purified water.

- **Advantages**: Glass is completely non-reactive, preserving water quality over long periods without leaching chemicals or absorbing odors.
- **Disadvantages**: Glass is heavy, prone to breakage, and impractical for storing large volumes of water. It also requires careful handling and is best stored in protected, stable environments.

3. Stainless Steel Containers

- **Stainless Steel Tanks and Jugs**: Stainless steel is another excellent material for water storage, especially for those concerned about the long-term durability and cleanliness of their containers. Stainless steel tanks are often used in professional settings, but they are also available for home use in sizes ranging from small jugs to large cisterns.
- **Advantages**: Stainless steel is corrosion-resistant, does not leach chemicals, and is highly durable. It also protects water from sunlight, which can prevent the growth of algae and bacteria.
- **Disadvantages**: Stainless steel containers are more expensive than plastic or glass alternatives, and they are heavier, making them more difficult to transport or move. Additionally, stainless steel may not be compatible with certain chemical water treatments like chlorine, as it can corrode over time.

4. Collapsible Water Containers

- **Portable Containers**: Collapsible water containers are made from flexible, food-grade plastic materials and are designed for portability and temporary storage. These containers are useful for emergency situations or for individuals who may need to transport water from one location to another.
- **Advantages**: Lightweight, portable, and compact when not in use. These containers are ideal for short-term water storage or emergency evacuation kits.
- **Disadvantages**: Collapsible containers are not as durable as rigid plastic or stainless steel options and are more prone to punctures or leaks. They are not recommended for long-term storage as they can degrade over time and may not provide the same level of protection against contaminants.

Avoid Non-Food-Grade Containers

Using non-food-grade containers for water storage can pose serious health risks. These containers may leach harmful chemicals into the water or harbor bacteria that can contaminate your supply. It's important to avoid containers that were previously used for non-food substances, such as:

- **Chemical Containers**: Never store water in containers that were previously used to hold chemicals, solvents, or paints. Even thorough cleaning cannot guarantee that all traces of harmful substances have been removed.
- **Plastic Containers Not Labeled as Food-Grade**: Many plastic containers, especially those used for industrial purposes, are not made from food-grade materials and may contain chemicals like BPA that can leach into the water. Always check for the food-grade label or the recycling code **#2 (HDPE)**, which is safe for water storage.

Container Size and Capacity

When choosing a container, consider the amount of water you need to store and how easily the containers can be moved or managed. Larger containers, like 55-gallon barrels, are ideal for long-term, stationary storage, while smaller jugs and bottles are more portable and easier to handle.

- **Small Containers (1-5 gallons)**: Ideal for daily use, smaller containers can be used for storing drinking water in accessible locations. They are portable, making them suitable for emergencies or transport.
- **Medium Containers (10-50 gallons)**: These containers can store a moderate amount of water for household needs without taking up too much space. They are easier to manage than larger barrels but still require proper storage space.
- **Large Containers (55 gallons or more)**: Large barrels and tanks are excellent for long-term storage and are often used in conjunction with rainwater harvesting systems. They are best suited for households with sufficient space and a dedicated water storage area.

Key Takeaways for Choosing Containers:

- Always use **food-grade containers** made from BPA-free plastic, glass, or stainless steel.
- **Plastic barrels** and tanks are ideal for long-term storage, but they must be kept in cool, dark locations to prevent degradation.
- **Glass** and **stainless steel** are highly durable and non-reactive but may not be practical for storing large volumes of water.
- Avoid any container that previously held **chemicals** or **non-food items** to prevent contamination.

By choosing the right containers, you can ensure that your stored water remains safe, clean, and available for long-term use, providing peace of mind in an off-grid environment.

Water Storage Locations and Conditions

Where and how you store your water is just as important as the container you use. Improper storage conditions can lead to contamination, degradation of the container, or the growth of harmful microorganisms like bacteria and algae. For long-term off-grid water storage, it is essential to find the right location and maintain optimal conditions to preserve the quality and safety of your water supply.

Ideal Conditions for Water Storage

1. Keep Water in a Cool, Stable Temperature

- **Temperature Range**: Store your water in an area with a consistent temperature, ideally between **50°F and 70°F (10°C - 21°C)**. Extreme temperatures can compromise the safety of your water. Excessive heat can speed up the growth of bacteria and algae, while freezing temperatures can cause containers to crack or burst, leading to contamination or loss of water.
- **Avoid Freezing Temperatures**: In cold climates, consider storing your water in insulated areas such as basements, underground storage, or inside your home. Water expands as it freezes, which can cause containers to rupture, making it important to avoid areas where temperatures drop below freezing.

2. Keep Water Away from Direct Sunlight

- **Sunlight Exposure**: Direct exposure to sunlight promotes the growth of algae, bacteria, and other harmful microorganisms in your stored water. Even clear, treated water can become contaminated if left in the sun for long periods. Additionally, UV rays from sunlight can degrade plastic containers, making them more prone to leaching chemicals into the water or breaking down over time.

- **Use Dark or Opaque Containers**: If storing water outdoors or in areas exposed to light, consider using dark-colored or opaque containers to block sunlight. If your containers are clear or translucent, make sure to store them in a shaded or covered area.

3. Elevate Containers Off the Ground

- **Protection from Contaminants**: Storing water directly on the ground can expose your containers to dirt, bacteria, and potential chemical spills. It's also more vulnerable to pests or flooding, which can lead to contamination. Place your containers on a platform, pallet, or shelf to keep them elevated and away from potential contaminants.
- **Prevent Damage**: Elevating containers also prevents wear and tear on the container's base. Continuous contact with rough or damp surfaces can weaken the container's structure over time, increasing the risk of leaks.

Choosing the Right Location for Water Storage

The location you choose for water storage can make a big difference in maintaining water quality. Different areas of your property or home have different environmental factors that may affect your stored water.

1. Indoor Storage

- **Basements or Cellars**: These areas are typically cool, dark, and stable, making them ideal for long-term water storage. Basements also offer protection from the elements, preventing exposure to temperature extremes and direct sunlight.
- **Utility Rooms or Closets**: If you don't have a basement, a utility room, pantry, or closet can be used for smaller quantities of water. These locations often provide easy access to your water supply in case of emergencies.
- **Avoid Attics**: While it may be tempting to store water in unused attic space, attics often experience high heat during summer months and may freeze in the winter, making them unsuitable for long-term water storage.

2. Outdoor Storage

- **Sheds or Garages**: If indoor space is limited, water can be stored in a garage or shed. However, these areas must be insulated to prevent freezing in winter and overheating in summer. Use insulated covers or wraps to protect your water barrels or tanks in these environments.
- **Buried Cisterns or Tanks**: For larger water storage needs, consider installing buried cisterns or underground tanks. These systems are naturally insulated by the earth, protecting your water from extreme temperatures and sunlight. They are ideal for rainwater harvesting systems or storing large volumes of water for household use.
- **Rainwater Harvesting Systems**: If you are storing water collected from rainwater harvesting, make sure the tanks or barrels are located in a shaded or sheltered area. Regularly check the system for leaks, clogs, or contamination risks, and ensure the storage tanks are sealed properly.

Avoiding Contamination Risks

1. Store Water Away from Chemicals and Fuels

- **Chemical Contamination**: It's essential to keep your water storage containers far away from potential sources of contamination such as chemicals, fertilizers, pesticides, gasoline, or other hazardous materials. Even small spills or fumes can permeate plastic containers and contaminate your water supply.
- **Separate Storage Areas**: If possible, dedicate a specific storage area just for water, away from cleaning supplies, fuels, or any other substances that could harm the integrity of your water.

2. Protect Against Flooding

- **Waterproof Storage**: If you live in an area prone to flooding, make sure your water storage containers are waterproof and elevated off the ground to prevent them from being submerged in floodwaters. Contaminated floodwater can seep into improperly sealed containers, rendering your stored water unsafe to drink.
- **Choose Higher Ground**: For outdoor water storage, always choose elevated areas that are less likely to flood. Avoid low-lying areas where rain runoff or standing water could collect around your containers.

3. Pest Prevention

- **Seal Containers Tightly**: Make sure that all containers are properly sealed with tight-fitting lids. Rodents, insects, and other pests are attracted to water, especially in dry areas where natural water sources are scarce. Sealed containers help protect your water from contamination and prevent pests from accessing your supply.

Best Practices for Long-Term Storage

- **Inspect Regularly**: Periodically check your water storage containers for signs of damage, leaks, or contamination. Look for cracks in plastic containers, rust on metal ones, and ensure that lids and seals remain tight.
- **Clean Storage Area**: Keep your water storage area clean, dry, and free of debris. A clean storage environment helps reduce the risk of contamination and makes it easier to detect potential problems early.
- **Monitor Temperature and Humidity**: If possible, use a thermometer or temperature monitor in your storage area to ensure that conditions remain stable. High humidity can lead to mold growth or rusting, which could affect water quality or the integrity of your containers.

Key Takeaways:

- Store water in a **cool**, **dark**, and **stable** environment to prevent bacterial growth and maintain water quality.
- Keep water containers **elevated** off the ground to protect them from contamination, pests, and potential damage.
- **Avoid direct sunlight** and **extreme temperatures** by using shaded areas, insulated covers, or underground storage options.
- Ensure your water storage area is **clean** and **free from chemicals** and hazardous materials to avoid contamination risks.

By following these best practices, you can ensure that your stored water remains safe and ready for use, even in long-term off-grid scenarios. Proper storage conditions are vital to maintaining the quality of your water supply and ensuring that it remains free from contamination.

Treating Water for Long-Term Storage

Even when water has been properly filtered and purified, it's essential to treat it before storing it for extended periods to prevent contamination and ensure it remains safe to drink. Over time, bacteria, algae, and other microorganisms can develop in stored water if not treated properly. To avoid this, using appropriate treatments will help preserve the water's quality and extend its shelf life.

This section explores various methods to treat water for long-term storage, including chemical treatments, preservatives, and best practices for ensuring that stored water remains potable.

1.4.3.1 Chlorine Treatment

Chlorine is one of the most effective and widely used methods for disinfecting water and preventing the growth of harmful microorganisms. By adding chlorine in the correct dosage, you can ensure that your water stays free of bacteria, viruses, and algae during long-term storage.

How to Use Chlorine to Treat Water for Storage

- **Chlorine Bleach**: Regular unscented household bleach (5-6% sodium hypochlorite) can be used to treat water. Ensure that the bleach is free from perfumes, dyes, and other additives.
- **Dosage**:
  - For **55-gallon barrels**: Add **½ teaspoon (about 40 drops)** of chlorine bleach per gallon of water. For a full 55-gallon barrel, this equates to approximately **2½ teaspoons**.
  - For **smaller containers**: Add **8 drops of bleach per gallon** of water.
- **Mix Well**: After adding the bleach, stir or shake the container to ensure the chlorine is evenly distributed throughout the water.
- **Waiting Period**: Allow the treated water to sit for at least **30 minutes** before use. After this period, the chlorine will have disinfected the water, making it safe for storage. There may be a faint chlorine odor, but this indicates that the water has been properly treated.
- **Shelf Life**: Water treated with chlorine can be stored for up to **6 months** before it should be rotated or re-treated. Over time, the chlorine will dissipate, so regular maintenance and rotation are essential.

Advantages of Chlorine Treatment

- **Highly Effective**: Chlorine kills most waterborne pathogens, making it a reliable method for long-term water disinfection.
- **Readily Available**: Household bleach is affordable and easy to find, making it a practical solution for off-grid water treatment.

Disadvantages

- **Taste and Smell**: Some people may find the taste or odor of chlorine-treated water unpleasant. However, this can be minimized by aerating the water (pouring it back and forth between containers) before drinking.
- **Not Effective Against All Contaminants**: Chlorine does not remove physical impurities, heavy metals, or chemical pollutants, so pre-filtration is essential.

Water Preserver Concentrates

Water preserver concentrates are commercially available solutions specifically designed for long-term water storage. These products are formulated to kill bacteria, viruses, mold, and algae while preserving the water's quality for several years. They are ideal for those who want a more advanced, low-maintenance solution for storing large quantities of water.

How to Use Water Preserver Concentrates

- **Follow Manufacturer Instructions**: Water preserver concentrates typically come with detailed instructions on dosage, which varies depending on the container size. Usually, a few drops or a small measured amount of the concentrate is added to each gallon of water.
- **Shelf Life**: Unlike chlorine, which needs to be reapplied every 6 months, water treated with a preserver concentrate can be stored for up to **5 years** without the need for re-treatment.
- **Application**: Water preserver concentrates can be used in any size container, from small jugs to large barrels and cisterns. Once added, the water should be sealed tightly to prevent contamination.

Advantages of Water Preserver Concentrates

- **Long Shelf Life**: Provides peace of mind by extending the safe storage time of water to several years without needing frequent rotation.
- **Tasteless and Odorless**: Unlike chlorine, water treated with preservatives typically does not have a noticeable taste or smell.

Disadvantages

- **Cost**: Water preserver concentrates are more expensive than household bleach and may be less accessible in some regions.
- **Limited Availability**: Not as widely available as chlorine or iodine treatments.

Other Chemical Treatments

Other chemicals, such as iodine and hydrogen peroxide, can also be used to treat water for long-term storage, though they are less common than chlorine or water preservatives. These treatments are effective but may come with specific advantages and disadvantages depending on the circumstances.

Iodine

- **How It Works**: Iodine is a powerful disinfectant and is commonly used in emergency water purification tablets. It can kill bacteria, viruses, and some parasites.
- **Dosage**: Iodine is typically used in tablet form or as a liquid solution. One tablet or a few drops of iodine solution is added per gallon of water, following the manufacturer's instructions.
- **Advantages**: Iodine is effective and easy to carry, making it a good option for short-term emergency storage or portable water kits.
- **Disadvantages**: Iodine can leave a strong taste and odor, which some people find unpleasant. It's also not recommended for long-term use, especially for individuals with thyroid conditions or those who are pregnant.

Hydrogen Peroxide

- **How It Works**: Hydrogen peroxide is another disinfectant that can be used to purify water. Food-grade hydrogen peroxide (35%) is preferred for water treatment, as it is highly concentrated and effective in small amounts.
- **Dosage**: Only a small amount (around **10 drops per gallon**) is required to treat water.
- **Advantages**: It is effective against bacteria and viruses, and like chlorine, it dissipates over time, leaving no residual taste or odor.
- **Disadvantages**: Hydrogen peroxide must be handled carefully, especially in its concentrated form, as it can be dangerous if ingested in high amounts. It's also less commonly used than chlorine or iodine for long-term water storage.

Best Practices for Treated Water Storage

1. Seal Containers Properly

Once water has been treated with chlorine, water preservatives, or other chemicals, make sure the containers are tightly sealed. This prevents contamination from the environment and keeps out pests or debris.

2. Store in a Cool, Dark Place

As with untreated water, treated water should be stored in a cool, dark location to avoid exposure to sunlight, heat, and temperature fluctuations. Sunlight can degrade some chemical treatments, particularly chlorine, reducing their effectiveness over time.

3. Rotate Water Regularly

Even when water has been treated for long-term storage, it's a good idea to rotate your water supply every **6 to 12 months**, unless you're using a preservative that guarantees a longer shelf life (up to 5 years). Always use the oldest water first and refill containers as needed.

4. Monitor for Contamination

Inspect your water storage regularly for any signs of contamination, such as cloudiness, sediment buildup, or foul odors. If the water appears contaminated, it should be re-filtered and treated before use.

Key Takeaways:

- **Chlorine treatment** is an affordable and effective method for preventing bacterial growth in stored water, but it may need reapplication every **6 months**.
- **Water preserver concentrates** offer longer storage solutions (up to **5 years**) without the need for frequent water rotation, but they are more expensive.
- **Iodine** and **hydrogen peroxide** are alternative chemical treatments, though less commonly used for long-term storage.
- Properly sealing and storing treated water in **cool, dark conditions** is essential to maintaining its quality over time.

By using the appropriate water treatments and following these best practices, you can ensure that your stored water remains safe and potable, providing security and peace of mind for long-term off-grid living or emergency preparedness.

4. Foundation of Off-Grid Power

In any off-grid living situation, energy efficiency is the first and most critical step to ensuring that you can sustainably meet your power needs. Whether you are generating electricity through solar, wind, or another off-grid power source, reducing your overall energy consumption will make your systems more reliable and easier to manage. Before you install any off-grid power source, it's essential to understand how to reduce your energy demand, as this will minimize the size and cost of your energy generation system.

This section will focus on the best practices for improving energy efficiency, reducing waste, and making the most of your off-grid power setup. By implementing these strategies, you can maximize the effectiveness of your power generation while minimizing the resources needed to maintain it.

Why Energy Efficiency is Essential in Off-Grid Living

When you are living off-grid, every watt of power counts. Unlike on-grid homes that have access to an unlimited power supply, off-grid systems are limited by the energy they can generate and store. The more energy efficient your home is, the less power you need to generate and store, making your system more cost-effective and easier to maintain.

Key Benefits of Energy Efficiency:

- **Reduced Energy Demand**: By minimizing energy consumption, you can reduce the size and complexity of your off-grid system, saving you money on initial setup and ongoing maintenance.
- **Extended Battery Life**: Lower energy consumption means that your battery storage lasts longer, providing more consistent and reliable power throughout the day and night.
- **Lower Environmental Impact**: By using energy-efficient appliances and reducing waste, you lower your overall carbon footprint, making your off-grid lifestyle more sustainable.

4.1 Assessing Your Energy Needs

Before setting up an off-grid power system, it's essential to first assess your household's energy needs. Understanding how much electricity you use daily will help you design a power system that is appropriately sized for your consumption patterns. This assessment will also highlight areas where you can reduce energy usage to make your system more efficient and cost-effective.

In this section, you'll learn how to evaluate your current energy consumption, calculate your daily energy needs, and identify the most power-hungry appliances in your home. This foundational step will help you tailor your off-grid energy system to your specific requirements, ensuring that it provides enough power without unnecessary overproduction or waste.

Steps to Assess Your Energy Usage

1. List Essential Appliances and Devices

Start by listing all the essential appliances and devices that you use on a daily basis. This includes not only large appliances but also smaller devices that contribute to your overall energy consumption. Consider everything from your refrigerator and water pump to lighting, laptops, and chargers.

Common appliances to include:

- Refrigerator or freezer
- Lighting (LEDs, CFLs, etc.)
- Water pump
- Laptop or computer
- Mobile phone chargers
- Cooking appliances (stove, microwave, etc.)
- Heating or cooling systems (fans, air conditioners, wood stoves)
- Entertainment devices (television, radio, etc.)

2. Determine the Power Consumption of Each Appliance

For each appliance on your list, find the power consumption, typically measured in **watts** (W). This information is usually found on the appliance's label or in the user manual. If the wattage is not readily available, you can often find this information online by searching for the appliance's model number and specifications.

If the appliance's power consumption is listed in **amperes (amps)**, you can convert it to watts using the following formula:

Watts (W)=Amps (A)×Volts (V)\text{Watts (W)} = \text{Amps (A)} \times \text{Volts (V)}Watts (W)=Amps (A)×Volts (V)

For example, if an appliance runs at **2 amps** and operates on **120 volts**, the power consumption is:

2A×120V=240W2A \times 120V = 240W2A×120V=240W

3. Calculate Daily Energy Use (Watt-Hours)

Once you know the wattage of each appliance, estimate how many hours per day you use it. Multiply the wattage by the number of hours to calculate the total daily energy consumption for each appliance, measured in **watt-hours (Wh)**.

Formula:

Watt-hours (Wh)=Watts (W)×Hours of Use (h)\text{Watt-hours (Wh)} = \text{Watts (W)} \times \text{Hours of Use (h)}Watt-hours (Wh)=Watts (W)×Hours of Use (h)

For example:

- If a **100-watt** refrigerator runs for **10 hours** a day, the daily consumption is: 100W×10h=1000Wh (or 1kWh)100W \times 10h = 1000Wh \text{ (or 1kWh)}100W×10h=1000Wh (or 1kWh)

4. Total Your Daily Energy Needs

After calculating the daily energy consumption for each appliance, add up the watt-hours for all appliances to find the total energy usage per day. This figure represents your daily energy needs and will be used to size your off-grid power system.

Example:

Appliance	**Power (W)**	**Usage (Hours/Day)**	**Daily Energy Use (Wh)**
Refrigerator	100W	10h	1000Wh
LED Lighting (5 bulbs)	50W	6h	300Wh
Laptop	50W	4h	200Wh
Water Pump	200W	2h	400Wh
Mobile Charger (2 units)	10W	2h	20Wh
Total Daily Energy Use			**1920Wh (1.92 kWh)**

In this example, the total energy consumption is **1.92 kWh/day**.

Estimating Seasonal Variations

It's important to remember that energy usage can vary depending on the season. For example, during winter months, you may use more power for heating and lighting, while in the summer, cooling appliances like fans or air conditioners may increase your energy demand. When assessing your energy needs, consider these seasonal variations and factor them into your calculations.

Adjustments for Seasonal Energy Use:

- **Winter**: More lighting, heating systems, and possibly extra power for hot water.
- **Summer**: Increased use of cooling systems like fans or air conditioners.

Identifying High-Energy Appliances

Certain appliances are known for their high energy consumption, making them significant contributors to your overall power usage. These appliances should be carefully considered when designing your off-grid system, as they may require additional power generation or storage capacity.

High-energy appliances include:

- **Refrigerators/Freezers**: These appliances run continuously and often consume the most energy in an off-grid home.
- **Air Conditioners and Heaters**: Heating and cooling systems can be major energy drains. Energy-efficient alternatives like wood stoves or passive cooling methods can reduce this burden.
- **Electric Stoves and Ovens**: Cooking with electricity can consume large amounts of energy. Off-grid homes often switch to gas stoves or outdoor cooking methods like solar ovens.

Optimizing Your Energy Usage

Once you have assessed your current energy needs, you can begin looking for ways to reduce consumption. This step will not only lower your overall power requirements but also reduce the cost and complexity of your off-grid power system. Simple changes like switching to energy-efficient appliances and adopting energy-saving habits can lead to significant reductions in daily energy use.

Energy-Saving Tips:

- **Switch to LED Lighting**: LED bulbs use up to 75% less energy than incandescent bulbs and last much longer, reducing your power consumption for lighting.
- **Use Energy-Efficient Appliances**: Replace old, inefficient appliances with modern, energy-saving models.
- **Unplug Devices**: Eliminate phantom loads by unplugging electronics when not in use or using smart power strips to shut off multiple devices at once.

Key Takeaways:

- **Assess your energy needs** by listing all essential appliances and calculating their daily power consumption in watt-hours.
- Identify **high-energy appliances**, such as refrigerators and air conditioners, and consider energy-efficient alternatives.
- **Total your daily energy usage** to estimate the size of the off-grid power system you need to install.

- **Seasonal variations** in energy demand should be accounted for to ensure your system can meet peak needs in winter or summer.
- **Reduce your energy consumption** wherever possible by switching to efficient appliances and adopting energy-saving habits.

By accurately assessing your energy needs, you can design an off-grid power system that provides sufficient electricity while minimizing costs and energy waste. This foundational step is crucial for ensuring a smooth transition to off-grid living and sustainable energy use.

4.2 Reducing Energy Consumption

Once you've assessed your energy needs, the next critical step in setting up an efficient off-grid power system is to reduce your overall energy consumption. By minimizing the amount of energy your household requires, you can reduce the size (and cost) of your power generation system and extend the life of your batteries and other energy storage components. This will also make your off-grid lifestyle more sustainable and easier to manage in the long run.

This section outlines practical strategies to reduce energy consumption in your home, focusing on upgrading appliances, improving home insulation, and adopting energy-efficient habits.

Why Reducing Energy Consumption is Important

In an off-grid environment, the amount of energy you consume directly impacts the size of the system you need to generate and store power. Every watt saved reduces the burden on your solar panels, wind turbines, batteries, and other off-grid power sources. Efficient energy use also makes your system more resilient, allowing it to handle periods of low energy production (e.g., during cloudy days or calm wind conditions).

Benefits of Reducing Energy Consumption:

- **Lower System Costs**: The smaller your energy demand, the fewer solar panels, batteries, and other system components you'll need, reducing the initial investment.
- **Longer Battery Life**: Reducing consumption extends the time your batteries can supply power, making your system more reliable.
- **More Sustainable**: Using less energy reduces your overall environmental footprint and makes your off-grid setup more sustainable.

Strategies for Reducing Energy Consumption

1. Upgrade to Energy-Efficient Appliances

One of the most effective ways to reduce energy consumption is by replacing old, inefficient appliances with modern, energy-efficient models. These appliances use significantly less power while delivering the same (or better) performance.

- **Energy Star Appliances**: Look for appliances with the **Energy Star** label, which certifies that the appliance meets energy efficiency standards set by the U.S. Environmental Protection Agency (EPA). Energy Star appliances use less electricity than standard models, which is particularly important for large, continuously running appliances like refrigerators and freezers.

Key upgrades to consider:

  - **Refrigerators and Freezers**: Energy-efficient refrigerators can reduce your power consumption by up to **30%** compared to older models. Consider opting for **DC-powered refrigerators** designed for off-grid use, as they work directly with solar or battery systems without requiring an inverter.
  - **LED Lighting**: Switch from incandescent or CFL bulbs to **LED bulbs**, which use up to **75% less energy** and last much longer. LED lighting is an essential upgrade for off-grid homes due to its low energy usage and durability.
  - **Washing Machines**: Choose energy-efficient washing machines that use less water and power. Front-loading machines tend to be more efficient than top-loaders.

---

2. Eliminate Phantom Loads

Phantom loads, also known as **vampire loads**, refer to the electricity consumed by devices when they are turned off but still plugged in. Common culprits include televisions, gaming consoles, chargers, and appliances with digital displays or standby modes.

- **Unplug Devices**: Unplug devices when they are not in use to prevent them from drawing power unnecessarily. For convenience, use **smart power strips** that allow you to easily cut off power to multiple devices at once.
- **Use Energy Meters**: Install **energy meters** to monitor phantom loads in your home. This can help you identify devices that are drawing power even when turned off, so you can take steps to eliminate waste.

---

3. Optimize Heating and Cooling

Heating and cooling systems are often the largest energy consumers in any home, especially in extreme climates. Reducing the need for artificial heating and cooling can have a significant impact on your overall energy usage.

- **Improve Home Insulation**: Proper insulation helps keep your home warm in winter and cool in summer, reducing the need for heating or air conditioning. Focus on insulating walls, floors, attics, and windows to prevent heat loss in the winter and heat gain in the summer.

Key areas to insulate:

  - **Windows and Doors**: Use double-glazed windows and weatherstripping to seal gaps around doors and windows, preventing drafts and reducing energy loss.
  - **Walls and Roofs**: Insulate exterior walls and the roof to improve your home's thermal efficiency.

- **Use Passive Solar Heating**: Design or modify your home to take advantage of **passive solar heating**, where sunlight naturally warms the interior. This involves using large south-facing windows to allow sunlight to enter during the day, and thermal mass materials (such as concrete or brick) to absorb and store heat for use at night.
- **Wood Stoves and Solar Heating**: Consider installing a **wood stove** or **solar water heater** to reduce your reliance on electric or gas heating. These options are ideal for off-grid living because they use renewable resources to heat your home and water.
- **Fans and Natural Ventilation**: In hot climates, reduce the need for air conditioning by using ceiling fans, which consume far less energy. You can also improve natural ventilation by opening windows and using vents to encourage airflow through your home.

---

4. Adopt Energy-Efficient Habits

Small changes in how you use energy can add up to significant savings. By adopting energy-efficient habits, you can reduce your overall consumption without the need for costly upgrades.

- **Use Appliances Wisely**:
  - **Run appliances during peak energy production**: If you are using solar panels, try to run high-energy appliances like washing machines, dishwashers, or vacuum cleaners during daylight hours when solar energy is most abundant.
  - **Batch cooking**: Cook multiple meals at once to make efficient use of your stove or oven, minimizing the energy used for heating.
- **Turn Off Lights**: Turn off lights in rooms that are not in use. Use **motion sensors** or **timers** to automatically turn off lights when they aren't needed.
- **Line-Dry Clothes**: Instead of using an electric dryer, opt for line-drying your clothes outdoors or indoors on a drying rack. Clothes dryers are energy-intensive, and using the sun or air to dry clothes is a free and sustainable alternative.
- **Use Cold Water for Laundry**: Wash clothes in cold water whenever possible, as heating water accounts for a large portion of the energy used in laundry.

---

5. Efficient Water Heating

Heating water is another significant energy expense in many households. By making your water heating system more efficient, you can reduce both your energy usage and your water consumption.

- **Solar Water Heaters**: Install a **solar water heater** to reduce or eliminate the need for electricity or gas to heat water. Solar water heaters use the sun's energy to heat water, making them an ideal solution for off-grid homes in sunny climates.
- **Insulate Water Pipes**: Insulating your hot water pipes reduces heat loss as water travels from the heater to your faucet, allowing you to use less energy to keep the water at the desired temperature.
- **Low-Flow Fixtures**: Install **low-flow faucets** and **showerheads** to reduce water usage, which in turn reduces the energy required to heat the water.

Key Takeaways:

- **Upgrade appliances** to energy-efficient models, such as Energy Star-rated refrigerators, LED lighting, and efficient washing machines.
- **Eliminate phantom loads** by unplugging devices when not in use and using smart power strips.
- **Optimize heating and cooling** through better insulation, passive solar heating, and the use of wood stoves or solar water heaters.
- **Adopt energy-efficient habits** such as turning off lights, line-drying clothes, and washing laundry with cold water to reduce daily energy consumption.

By following these strategies, you can significantly reduce your overall energy consumption, making your off-grid power system more manageable, cost-effective, and sustainable. Every watt saved is a step toward achieving energy independence and living more efficiently off the grid.

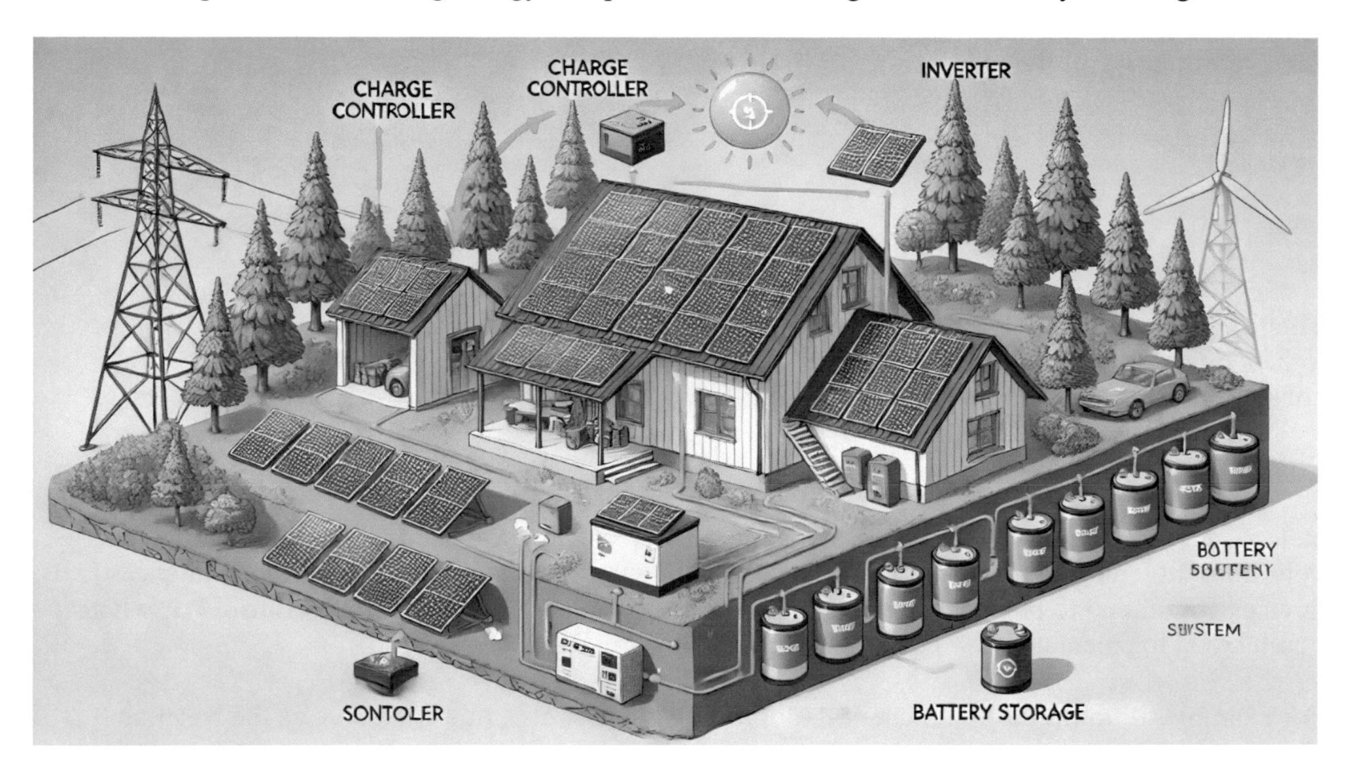

5. Solar Power for Sustainable Energy

Solar power is one of the most reliable and accessible energy sources for off-grid living. It allows you to generate electricity using sunlight, providing a sustainable and renewable way to power your home without relying on fossil fuels or utility grids. Setting up an off-grid solar power system can be a game-changer for those seeking energy independence, and it is essential to understand the components, system design, and considerations needed to make it work effectively.

This section will guide you through the basics of off-grid solar power, from calculating your energy requirements to choosing the right components and understanding how to set up and maintain the system. With careful planning, you can use solar energy to meet your daily power needs and achieve true off-grid living.

5.1 Determining Solar Requirements

The first step in setting up an off-grid solar system is determining how much solar power you need to generate to meet your energy demands. This is calculated based on your daily energy consumption and the amount of sunlight your location receives.

Step 1: Calculate Your Daily Energy Use

Before you can determine the size of your solar power system, you need to know how much energy your household consumes daily. This involves adding up the energy usage of all the appliances and devices you intend to power with solar energy. This was covered in detail in **Section 2.1.1**, where you learned to calculate daily energy use in **watt-hours (Wh)**.

For example, if your household uses **5 kWh** (5000 Wh) per day, this will serve as the baseline for sizing your solar system.

Step 2: Evaluate Your Solar Resource

The next factor to consider is the amount of sunlight your location receives, measured in **peak sun hours**. This varies depending on your geographical location, climate, and the time of year. A **peak sun hour** is equivalent to one hour of sunlight with an intensity of **1000 watts per square meter**. Most locations in the United States receive between **3 to 6 peak sun hours per day**.

You can find the average peak sun hours for your location by using online solar calculators or local data from meteorological services.

Step 3: Calculate the Solar Panel Size

Once you have your daily energy consumption and peak sun hours, you can calculate how much solar capacity you need. Divide your total daily energy use (in watt-hours) by the number of peak sun hours your location receives. This will give you the total wattage of solar panels required.

Formula:

Solar Panel Size (W)=Daily Energy Use (Wh)Peak Sun Hours (h)\text{Solar Panel Size (W)} = \frac{\text{Daily Energy Use (Wh)}}{\text{Peak Sun Hours (h)}}Solar Panel Size (W)=Peak Sun Hours (h)Daily Energy Use (Wh)

For example:

- If your daily energy use is **5000 Wh** and you receive **5 peak sun hours** per day, you would need a system capable of generating: 5000 Wh5 h=1000 watts of solar panels\frac{5000 \text{ Wh}}{5 \text{ h}} = 1000 \text{ watts of solar panels}5 h5000 Wh =1000 watts of solar panels

This means you would need approximately **1000 watts (1 kW)** of solar panels to meet your energy needs.

5.2 Types of Solar Panels and Battery Storage

There are several types of solar panels available for off-grid systems, each with its advantages and disadvantages. The most common types are **monocrystalline**, **polycrystalline**, and **thin-film** solar panels.

1. Monocrystalline Solar Panels

- **Efficiency**: These panels are made from high-purity silicon, which gives them a higher efficiency rate (around **17-22%**) compared to other types of panels. This means they can produce more power in a smaller area.
- **Advantages**:
    - Higher efficiency.
    - Longer lifespan (typically 25 years or more).
    - Perform better in low-light conditions.
- **Disadvantages**:
    - More expensive than polycrystalline and thin-film panels.
    - Slightly less efficient at very high temperatures.

2. Polycrystalline Solar Panels

- **Efficiency**: Polycrystalline panels are made from silicon fragments melted together, resulting in lower efficiency (around **15-17%**) compared to monocrystalline panels. They require more space to generate the same amount of power.
- **Advantages**:
    - More affordable than monocrystalline panels.
    - Suitable for larger installations where space is not an issue.
- **Disadvantages**:
    - Lower efficiency and performance in low-light conditions.
    - Slightly shorter lifespan than monocrystalline panels.

3. Thin-Film Solar Panels

- **Efficiency**: Thin-film panels are made from a variety of materials, including amorphous silicon, cadmium telluride, and copper indium gallium selenide. These panels have lower efficiency (around **10-12%**) but are lightweight and flexible.
- **Advantages**:
    - Flexible and can be applied to various surfaces (rooftops, walls, etc.).
    - Perform better in high-temperature environments.
- **Disadvantages**:
    - Low efficiency, requiring more space.
    - Shorter lifespan compared to crystalline panels.

Choosing the Right Solar Panels: For most off-grid setups, **monocrystalline** or **polycrystalline** panels are preferred due to their higher efficiency and durability. If space is limited, monocrystalline panels may be the better option. For larger installations where budget is a concern, polycrystalline panels may offer a more cost-effective solution.

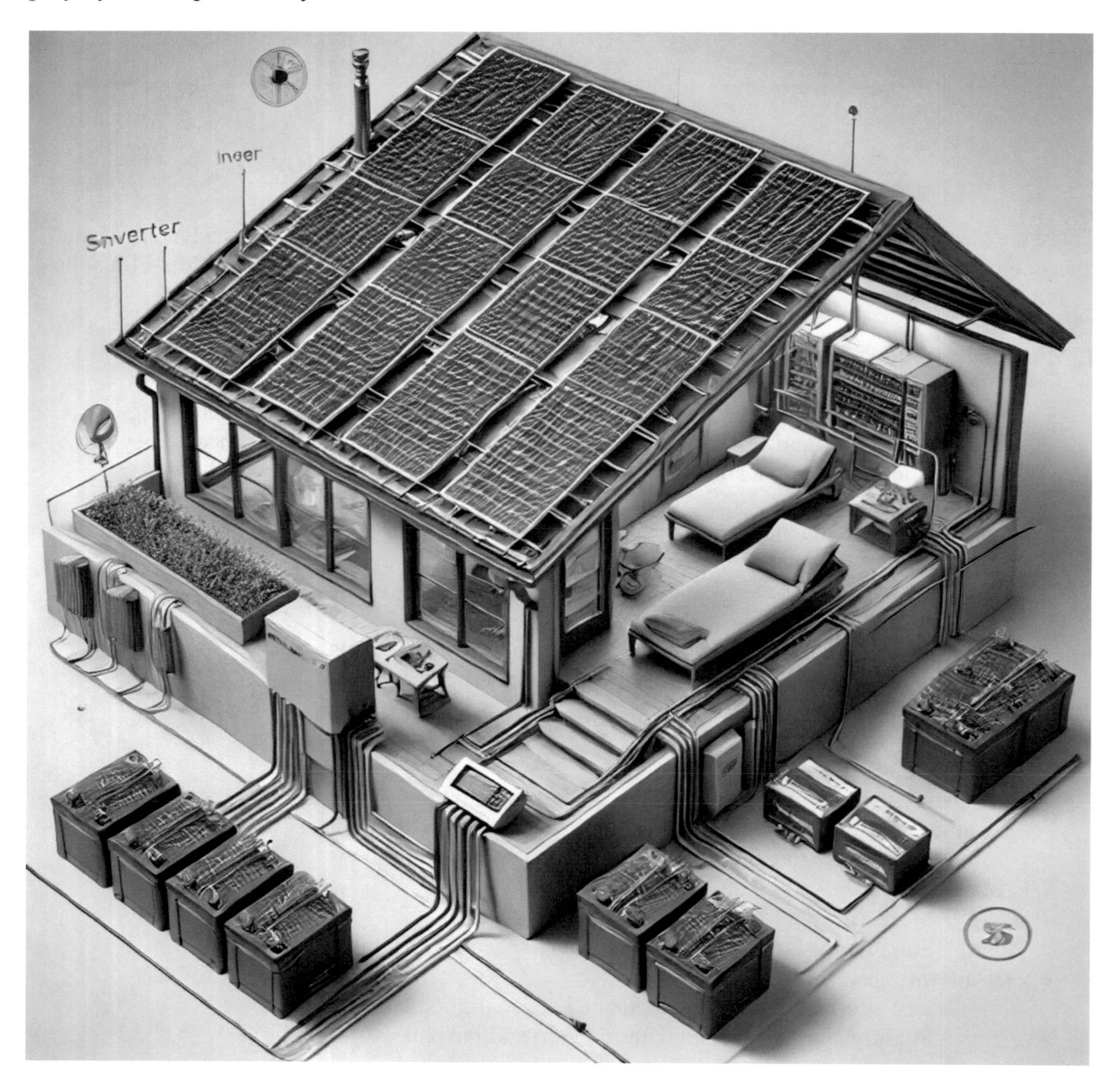

5.2.1 Battery Storage

In an off-grid solar system, batteries are essential for storing excess energy generated during the day for use at night or during periods of low sunlight. Choosing the right battery type and size is crucial for ensuring a reliable power supply.

Types of Batteries:

1. **Lead-Acid Batteries**:
    - **Flooded Lead-Acid (FLA)**: These are the most affordable batteries, but they require regular maintenance (checking and refilling water levels). They have a shorter lifespan compared to other battery types.
    - **Sealed Lead-Acid (SLA)**: These are maintenance-free and can be used indoors, but they are slightly more expensive than flooded lead-acid batteries.
2. **Lithium-Ion Batteries**:
    - **Advantages**: Lithium-ion batteries have a longer lifespan, higher efficiency, and greater depth of discharge compared to lead-acid batteries. They are also maintenance-free and have a smaller footprint.
    - **Disadvantages**: The main drawback is the higher upfront cost, but they typically pay off over time due to their longevity and performance.
3. **Saltwater Batteries**:
    - **Advantages**: These are environmentally friendly, non-toxic batteries that use saltwater as the electrolyte. They have a long lifespan and are fully recyclable.
    - **Disadvantages**: They are less efficient than lithium-ion batteries and relatively new to the market, so they may not be as widely available.

Choosing the Right Battery: **Lithium-ion** batteries are becoming the most popular choice for off-grid solar systems due to their efficiency and longer lifespan. However, **lead-acid** batteries are still widely used in budget-conscious setups. If environmental impact is a major concern, consider **saltwater batteries** as an alternative.

5.2.2 Solar Charge Controllers and Inverters

Solar Charge Controllers

The solar charge controller is a crucial component in any solar system. It regulates the voltage and current coming from the solar panels to prevent overcharging the batteries.

- **Types of Charge Controllers**:
    - **Pulse Width Modulation (PWM)**: These are more affordable but less efficient. They are best suited for smaller systems.
    - **Maximum Power Point Tracking (MPPT)**: These are more efficient and capable of handling higher voltages, making them ideal for larger systems. MPPT controllers can extract more power from the solar panels, especially in low-light conditions.

Inverters

Inverters are necessary to convert the **DC power** generated by your solar panels into **AC power**, which most household appliances use. There are different types of inverters, including:

- **Pure Sine Wave Inverters**: These are the most efficient and reliable for running sensitive electronics like computers and appliances.
- **Modified Sine Wave Inverters**: These are less expensive but may not work well with all appliances, particularly those with motors or sensitive electronics.

Key Takeaways:

- To determine your solar requirements, calculate your daily energy usage and divide it by the number of peak sun hours in your location.
- Choose the right solar panels based on efficiency, budget, and available space. **Monocrystalline** panels offer higher efficiency, while **polycrystalline** panels are more affordable.
- Invest in a battery storage system to ensure power availability when sunlight is limited. **Lithium-ion** batteries offer the best performance, but **lead-acid** and **saltwater** batteries may be better suited to specific budgets or environmental concerns.
- Use **MPPT solar charge controllers** for better efficiency and **pure sine wave inverters** to convert DC to AC power for your appliances.

By carefully planning and selecting the appropriate components, you can design an off-grid solar system that meets your energy needs, is cost-effective, and provides long-term sustainability for your off-grid lifestyle.

5.3 Advanced Solar Panel Configuration and Maintenance

As you progress in your off-grid energy journey, understanding advanced solar panel configurations and maintenance is crucial for maximizing efficiency and ensuring the longevity of your system. This section will dive into more complex setups, techniques for improving performance, and essential maintenance practices to keep your solar power system running smoothly, even in challenging conditions.

---

1. Parallel vs. Series Wiring: Maximizing Efficiency

Solar panels can be wired in **series** or **parallel**, and each configuration offers different benefits depending on your energy needs and environmental conditions. Here's how they work:

- **Series Wiring**: In this setup, the positive terminal of one panel is connected to the negative terminal of the next, increasing the system's overall voltage while keeping the current constant. This is beneficial when using long cables or when your inverter requires higher voltage input.
  - **Advantages**: Series wiring minimizes power loss over long distances and is ideal for setups where the panels receive even sunlight.

    - **Disadvantages**: A drop in performance from one panel (due to shading, dirt, or damage) can reduce the efficiency of the entire system, as the current is limited by the weakest panel.
- **Parallel Wiring**: In parallel wiring, the positive terminals of all panels are connected, and the negative terminals are also connected, keeping the voltage constant while increasing the current.
    - **Advantages**: This setup allows each panel to operate independently, so shading or damage to one panel doesn't affect the others.
    - **Disadvantages**: Higher currents require thicker, more expensive cables and can lead to higher energy losses over distance.

Choosing the Right Setup: If you live in an area with consistent sunlight and minimal shading, series wiring may offer the best efficiency. However, if shading is a concern or you expect inconsistent sun exposure, parallel wiring provides more reliability. You can also use a combination of series and parallel wiring for the best of both worlds.

Tip: Include a **diagram** illustrating both series and parallel wiring configurations to help readers visualize the setups.

---

2. Battery Bank Design and Maintenance

The efficiency of your solar system is closely tied to your battery bank, which stores the energy generated during the day for use at night or during cloudy weather. Here's how to optimize your battery system:

- **Battery Types**: Choose between **lead-acid batteries** (cheaper but require regular maintenance) and **lithium-ion batteries** (more expensive but longer-lasting and maintenance-free).
    - **Lead-Acid Batteries**: These require regular checks to ensure water levels are adequate, and they should not be discharged below 50% to maintain their lifespan.
    - **Lithium-Ion Batteries**: These are more efficient and last longer but come at a higher upfront cost. They can be discharged up to 80% without harming the battery.
- **Battery Maintenance**: Proper maintenance is critical for keeping your batteries in top condition:
    - **Temperature Control**: Batteries perform best at moderate temperatures. Use battery enclosures or insulated containers to protect them from extreme heat or cold.
    - **Regular Testing**: Test your battery voltage and specific gravity (for lead-acid batteries) every few months to ensure they are holding a charge properly.
    - **Equalization**: For lead-acid batteries, periodically perform an equalization charge to prevent sulfation (a condition where sulfate crystals build up on the battery plates, reducing capacity).

Tip: Add a **battery maintenance checklist** to ensure readers are aware of the key tasks they should perform to maintain their system's performance.

---

3. Maximizing Winter Performance: Solar System Optimization

During the winter months, shorter days and cloudy weather can reduce solar energy production. Here are some strategies to optimize your system during low-sunlight periods:

- **Adjusting the Tilt Angle**: In winter, the sun is lower in the sky. To capture the maximum amount of sunlight, adjust the tilt angle of your panels to a steeper angle (closer to vertical). For most northern locations, aim for a tilt angle of **60 to 70 degrees** in winter.
- **Clearing Snow and Debris**: Snow accumulation on panels can block sunlight. Regularly clear snow and debris from your panels to ensure they are fully exposed. A long-handled squeegee or brush is useful for this purpose.
- **Energy Conservation Strategies**: To conserve energy during low-production periods, reduce energy usage by prioritizing essential appliances. Consider adding a **backup generator** or increasing battery capacity to store more energy during sunny days.
- **Energy Efficiency**: During periods of lower energy production, maximizing efficiency is essential. Use energy-efficient appliances and consider installing LED lights and energy-efficient heating systems to reduce your overall power consumption.

---

4. Solar Panel Maintenance: Ensuring Longevity and Performance

Solar panels are relatively low maintenance, but periodic cleaning and inspection will help ensure they continue to operate at peak efficiency for many years.

- **Cleaning Your Panels**: Dust, bird droppings, and debris can block sunlight and reduce panel efficiency by up to 30%. Clean your panels regularly using a soft cloth, water, and a non-abrasive soap. Avoid using harsh chemicals or abrasive materials that could scratch the panels.
    - **Frequency**: Cleaning is especially important in dry, dusty environments or if you notice a drop in energy production. Aim to clean your panels at least twice a year, or more frequently in dusty regions.
- **Visual Inspection**: Inspect your panels for any signs of physical damage, such as cracks in the glass or damage to the wiring and connections. Even minor damage can reduce performance and may require professional repairs.
- **Check for Shading**: As trees grow or new structures are built, shading can occur where it wasn't an issue before. Periodically check for shading during peak sun hours and trim any overhanging branches that may block sunlight.

Tip: Include a **solar maintenance log** template where readers can track cleaning schedules, inspections, and any repairs made to their system.

---

5. Tools and Equipment for Advanced Solar System Monitoring

To keep your solar system running efficiently, it's important to monitor its performance regularly. Here are some advanced tools that can help:

- **Multimeter**: This tool measures the voltage, current, and resistance of your solar panels and battery bank. Regular checks with a multimeter will help ensure your system is producing the expected amount of energy.

- **Solar Power Meters**: These meters measure solar radiation and give you a clear indication of how much sunlight your panels are receiving. This is useful for diagnosing issues with panel placement or determining whether adjustments are needed.
- **Battery Monitors**: Installing a battery monitor helps you track battery charge levels, power consumption, and overall system efficiency. This real-time data allows you to manage energy consumption effectively, especially during low-production periods.

Key Takeaways

- **Choose the Right Wiring Configuration**: Series wiring increases voltage, while parallel wiring increases current. Select the configuration that best suits your energy needs and environmental conditions.
- **Maintain Your Battery Bank**: Whether you use lead-acid or lithium-ion batteries, regular maintenance is essential for maximizing the lifespan and performance of your system.
- **Optimize Winter Performance**: Adjust panel tilt, clear snow, and conserve energy during the winter months to keep your system running efficiently.
- **Regular Maintenance is Key**: Cleaning your panels and inspecting them for damage will ensure long-term performance. Keep track of maintenance with a solar maintenance log.

By implementing these advanced configurations and maintenance strategies, you can maximize the efficiency and lifespan of your solar system, ensuring a reliable and sustainable energy supply for your off-grid lifestyle.

6. Solar Power System Troubleshooting Guide

Even with a well-designed solar power system, issues can arise that affect performance and reliability. In this section, we'll cover common solar system problems, diagnostic tools, and step-by-step troubleshooting procedures to help you identify and fix issues quickly, ensuring your off-grid energy system runs smoothly.

---

1. Common Solar System Failures

Solar systems can encounter several types of failures, ranging from minor inefficiencies to significant malfunctions. Here are the most frequent issues you may face:

- **Low Energy Production**: If your system is generating less power than expected, this could be due to shading, dirt buildup on the panels, or faults in the wiring or inverters.
- **Battery Not Charging Properly**: Sometimes, your solar batteries may not charge fully, which could indicate problems with your charge controller, wiring, or the batteries themselves.
- **Inverter Malfunctions**: The inverter, which converts DC (direct current) power from the solar panels into usable AC (alternating current) power, can sometimes fail, leading to no power output to your appliances.
- **Overheating Components**: High temperatures can reduce the efficiency of solar panels, inverters, and batteries. Overheating can result from poor ventilation, incorrect setup, or faults in the system.

- **Inconsistent or Flickering Power**: Flickering or inconsistent power can be a sign of poor wiring, connection issues, or problems with the inverter.

---

2. Step-by-Step Troubleshooting Guide

When your solar system isn't functioning properly, follow this step-by-step guide to diagnose and fix common issues:

Step 1: Inspect the Solar Panels

- **Check for Shading**: The most common cause of low energy production is shading. Ensure that your panels are free from shade during peak sun hours. Overhanging trees, buildings, or even dirt can reduce efficiency. Trim any trees and clean the panels if necessary.
- **Clean the Panels**: Dust, bird droppings, and grime can accumulate on the surface of your solar panels, blocking sunlight. Clean the panels using a soft brush or cloth with water and a mild detergent. Avoid using abrasive materials.
- **Inspect for Physical Damage**: Look for cracks, chips, or damage to the solar cells. Even minor damage can significantly reduce efficiency. If you spot damage, contact a professional to replace the affected panels.

Step 2: Test the Battery Bank

- **Check Battery Voltage**: Using a multimeter, measure the voltage of your battery bank. Compare the voltage reading to the manufacturer's specifications. If the voltage is significantly lower than expected, this may indicate that the batteries are not charging properly.
    - **Lead-Acid Batteries**: For lead-acid batteries, a fully charged battery should have a voltage of around 12.6 to 12.8 volts (for 12V batteries).
    - **Lithium-Ion Batteries**: Lithium-ion batteries should have a voltage of 12.4 to 12.7 volts when fully charged.
- **Check Battery Connections**: Loose or corroded battery terminals can cause charging issues. Ensure that all connections are tight and clean, and remove any corrosion with a wire brush or baking soda solution.
- **Perform a Load Test**: A load test can help determine if the batteries are holding a charge properly. This involves applying a load to the battery bank and measuring how the voltage responds over time. If the voltage drops rapidly, the batteries may need to be replaced.

Step 3: Inspect the Charge Controller

- **Verify Controller Settings**: Check that your charge controller is set up correctly for your battery type (lead-acid, lithium-ion, etc.) and voltage. Misconfigured settings can prevent your batteries from charging properly.
- **Check Charge Controller Display**: Many charge controllers come with an LCD display or indicator lights that show the status of your solar system. If there are any error codes or warning lights, refer to the user manual for troubleshooting specific to your controller model.
- **Test Input and Output Voltage**: Use a multimeter to measure the voltage at the input (solar panel connection) and output (battery connection) of the charge controller. Ensure the

input voltage matches the solar panel's output and that the charge controller is delivering the correct voltage to the battery bank.

Step 4: Diagnose Inverter Issues

- **Check Inverter Display**: Like charge controllers, most inverters have a display that shows system performance and error codes. If there's an issue, the display will often provide a clue as to what's wrong (e.g., overload, overheating).
- **Test Input Voltage**: Measure the DC voltage going into the inverter from the battery bank. It should be consistent with the battery's voltage. If the voltage is too low or fluctuating, the batteries or wiring could be the issue.
- **Test AC Output Voltage**: If the inverter is functioning correctly, measure the AC voltage on the output side. For standard systems, this should be around 110V or 220V depending on your location. If the voltage is lower than expected or inconsistent, the inverter may need to be repaired or replaced.
- **Check for Overheating**: Inverters can overheat if not properly ventilated or if they are overworked. Ensure there is enough airflow around the inverter and check for dust buildup in the cooling vents. If overheating persists, consider upgrading to a higher-capacity inverter.

Step 5: Examine Wiring and Connections

- **Check for Loose or Corroded Connections**: Go through the entire system (panels, charge controller, inverter, battery bank) and inspect all wiring connections. Tighten any loose wires and clean any corrosion you find. Corroded or loose wires can cause power drops and inconsistent energy flow.
- **Look for Damaged Wires**: Rodents or weather damage can cause wires to fray or short circuit. Carefully inspect all cables for signs of wear, and replace any damaged wires immediately.
- **Verify Wire Gauge**: Ensure you are using the correct gauge wire for your system. Inadequate wire thickness can cause energy loss, overheating, and system inefficiencies, especially over long distances.

---

3. Tools for System Diagnosis

Having the right tools on hand makes troubleshooting much easier. Here are some key diagnostic tools you should keep in your off-grid toolbox:

- **Multimeter**: A multimeter is essential for testing voltage, current, and resistance throughout your solar system. Use it to check battery voltage, verify solar panel output, and diagnose inverter issues.
- **Infrared Thermometer**: Use this tool to detect overheating in components like batteries, inverters, and charge controllers. Overheating can indicate wiring issues or system overloading.
- **Solar Power Meter**: This measures solar radiation to help determine if your panels are receiving the expected amount of sunlight. It's particularly useful for diagnosing shading or tilt angle problems.

---

4. Preventive Maintenance Tips

To avoid many of the common issues described above, regular maintenance is key. Here are some best practices for keeping your system in top shape:

- **Panel Cleaning Schedule**: Clean your panels every 6 months or as needed, especially after heavy dust or pollen seasons. Keeping the panels clear of debris ensures maximum sunlight absorption.
- **Battery Checkups**: Inspect your battery bank monthly, checking for signs of wear, corrosion, or overheating. For lead-acid batteries, check water levels and top up with distilled water if necessary.
- **Inverter and Charge Controller Ventilation**: Ensure that your inverter and charge controller are placed in well-ventilated areas and that any fans or cooling mechanisms are functioning properly.
- **Wiring Inspections**: Every few months, inspect your system's wiring for signs of wear, damage, or looseness. Replace any frayed wires immediately to prevent energy loss or short circuits.

Key Takeaways for Solar Power Troubleshooting

- **Common Issues**: Low energy production, faulty batteries, malfunctioning inverters, and poor wiring are the most common problems.
- **Step-by-Step Diagnostics**: Start by inspecting the panels for shading and damage, then move through the charge controller, batteries, and wiring. Use a multimeter to test voltage at key points.
- **Preventive Maintenance**: Regularly cleaning panels, testing batteries, and inspecting wiring can prevent most system failures before they become serious.
- **Tools**: Keep essential tools like a multimeter and infrared thermometer in your toolkit to diagnose issues quickly and accurately.

By following this troubleshooting guide, you'll be able to identify and resolve most solar power system issues on your own, ensuring a steady and reliable energy supply for your off-grid lifestyle.

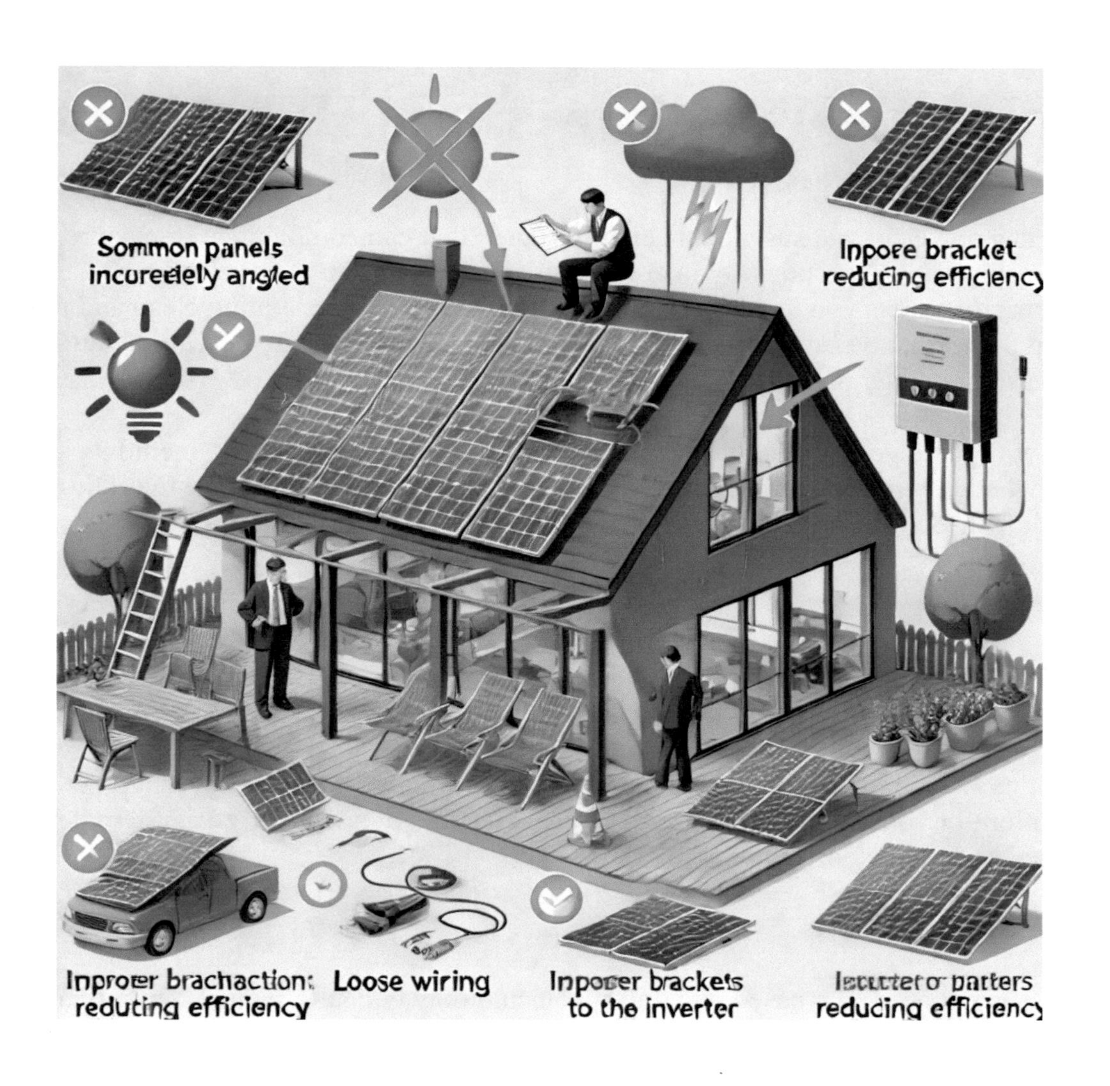
Sommon panels
incuredely angled
Inpore bracket
reducing efficiency
Inproer brachaction:
reduting efficiency
Loose wiring
Inpoer brackets
to the inverter
reducing efficiency

6. Exploring Wind Power and Hybrid Systems

Wind power is a powerful and sustainable energy source that can complement or even replace solar power in off-grid systems, particularly in areas with consistent wind. Using wind turbines to generate electricity allows you to take advantage of nature's power, providing clean and renewable energy. Wind power can be especially useful in regions with strong winds, during nighttime, or in seasons when solar energy production may be lower.

This section will guide you through the fundamentals of off-grid wind power, including determining whether wind is a viable energy source for your location, choosing the right wind turbine, and how to set up and maintain a wind power system.

6.1 Evaluating Wind Power Suitability

Before investing in a wind turbine, it's essential to assess whether wind energy is a practical and efficient solution for your specific location. The effectiveness of a wind power system depends on the wind speed and consistency in your area.

1. Wind Speed Requirements

Wind turbines generate power based on wind speed, and they typically require a minimum average wind speed of **7 to 10 miles per hour (mph)** to function efficiently. The higher the wind speed, the more energy a turbine can generate.

- **Wind Classifications**: Wind speeds are classified into different levels, with **Class 3** (12.5–14.3 mph) or higher being the most suitable for wind power systems. Lower wind classes (below 10 mph) may not generate enough power to justify the investment in a wind turbine.

2. Tools for Measuring Wind Speed

To determine if your location is suitable for wind power, you can measure the wind speed using:

- **Anemometers**: Handheld devices or installed units that measure wind speed over time.
- **Online Wind Maps**: Various wind resource maps and online tools provide regional data on average wind speeds based on geographical location.

3. Wind Consistency

Beyond wind speed, consistency is key. Ideal wind power locations experience steady winds year-round, without large fluctuations in speed or direction. While some variability is natural, areas with constant, moderate to strong winds will produce the most reliable power.

6.1.1 Types of Small Wind Turbines

Wind turbines come in different sizes and configurations, designed for various applications. For off-grid living, small-scale turbines (also known as **micro** or **small wind turbines**) are typically used. These turbines are suitable for residential properties, small farms, or cabins, and they can provide enough electricity to power essential household appliances.

1. Horizontal-Axis Wind Turbines (HAWT)

- **Description**: The most common type of wind turbine, horizontal-axis wind turbines have a propeller-like design, with blades that rotate around a horizontal axis.
- **Advantages**:
    - High efficiency in strong, consistent winds.
    - Suitable for larger power production.
- **Disadvantages**:
    - Requires significant space and height for installation.
    - Can be noisy, depending on the model.

2. Vertical-Axis Wind Turbines (VAWT)

- **Description**: Vertical-axis wind turbines have blades that rotate around a vertical axis, resembling a helix or egg beater. These turbines are less common but can be more practical for residential or small-scale applications.
- **Advantages**:
    - Can capture wind from any direction, making them suitable for areas where wind direction shifts frequently.
    - Operates well at lower wind speeds and can be installed closer to the ground.
- **Disadvantages**:
    - Generally less efficient than horizontal-axis turbines.
    - Requires more maintenance due to the vertical design.

Choosing the Right Turbine: For most off-grid homes, **horizontal-axis wind turbines** are the better choice if you have the space and access to strong winds. **Vertical-axis turbines** are a good option for smaller properties or areas with unpredictable wind patterns.

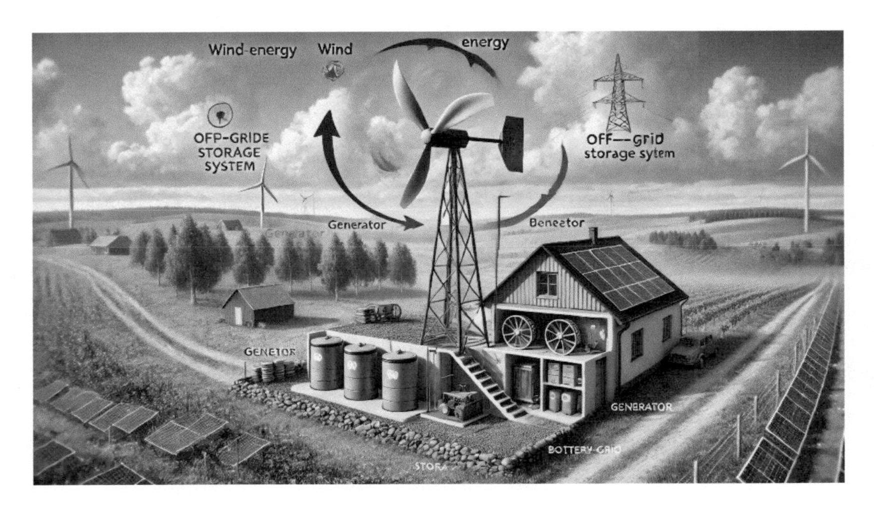

6.2 Setting Up Your DIY Wind Turbine Kit

If you're opting for a DIY wind power solution, wind turbine kits provide all the components you need to set up a functional system. These kits can be a cost-effective way to install a wind turbine on your property without needing professional installation services.

Components of a Wind Power System:

1. **Wind Turbine**: The main component that captures wind energy and converts it into electrical energy.
2. **Tower**: Wind turbines need to be elevated to catch sufficient wind. Towers can range from **30 to 100 feet** high, depending on your location and wind conditions.
3. **Charge Controller**: Regulates the power generated by the wind turbine to prevent overcharging your battery bank.
4. **Inverter**: Converts the DC power generated by the turbine into AC power, which most household appliances use.
5. **Batteries**: Stores excess power generated by the wind turbine for use when the wind is not blowing or energy demand is high.

Step-by-Step Wind Turbine Installation:

1. **Choose the Right Location**: Select an open area with minimal obstructions such as buildings or trees. The turbine should be placed at least **30 feet** higher than any surrounding obstacles.
2. **Assemble the Wind Turbine**: Follow the manufacturer's instructions to assemble the wind turbine. Most DIY kits come with detailed guides to help you through the process.
3. **Install the Tower**: Securely install the tower according to the kit's specifications. Towers may be guyed (supported by tensioned wires) or freestanding. Make sure the foundation is strong and stable to withstand wind forces.
4. **Connect the Electrical Components**: Connect the wind turbine to the charge controller, inverter, and battery bank. Ensure all connections are secure and weatherproofed.
5. **Test the System**: Once the system is installed, perform a test run to ensure the turbine is generating power and that all components are functioning correctly.

6.2.1 Installing a Small Wind Turbine

For those who prefer a pre-built solution, installing a small wind turbine involves many of the same steps as setting up a DIY kit. The main difference is that commercially available small turbines are typically more robust and may come with additional features for monitoring and adjusting the system.

Professional Installation:

In some cases, it's worth hiring professionals to install the wind turbine, particularly if the system is large or requires specialized equipment. Professional installation ensures that the turbine is correctly positioned, anchored, and connected to your off-grid system.

6.3 Combining Wind and Solar Power for Energy Independence

One of the best ways to maximize the reliability of your off-grid energy system is to combine wind and solar power. This creates a **hybrid system** that leverages the strengths of both energy sources, ensuring you have power available regardless of weather conditions.

Benefits of Hybrid Systems:

- **Complementary Energy Production**: Solar and wind power complement each other well. Solar energy is most abundant during sunny, calm days, while wind power can generate electricity during cloudy or windy days, often at night.
- **Increased Energy Security**: A hybrid system ensures that you have power available even if one energy source is temporarily unavailable due to weather conditions.

How to Set Up a Hybrid System:

- **Shared Battery Bank**: Both solar panels and wind turbines can be connected to the same battery bank, allowing excess energy from both sources to be stored for later use.
- **Dual Charge Controllers**: You'll need separate charge controllers for each power source (one for the solar panels and one for the wind turbine) to ensure both systems operate efficiently and without interference.

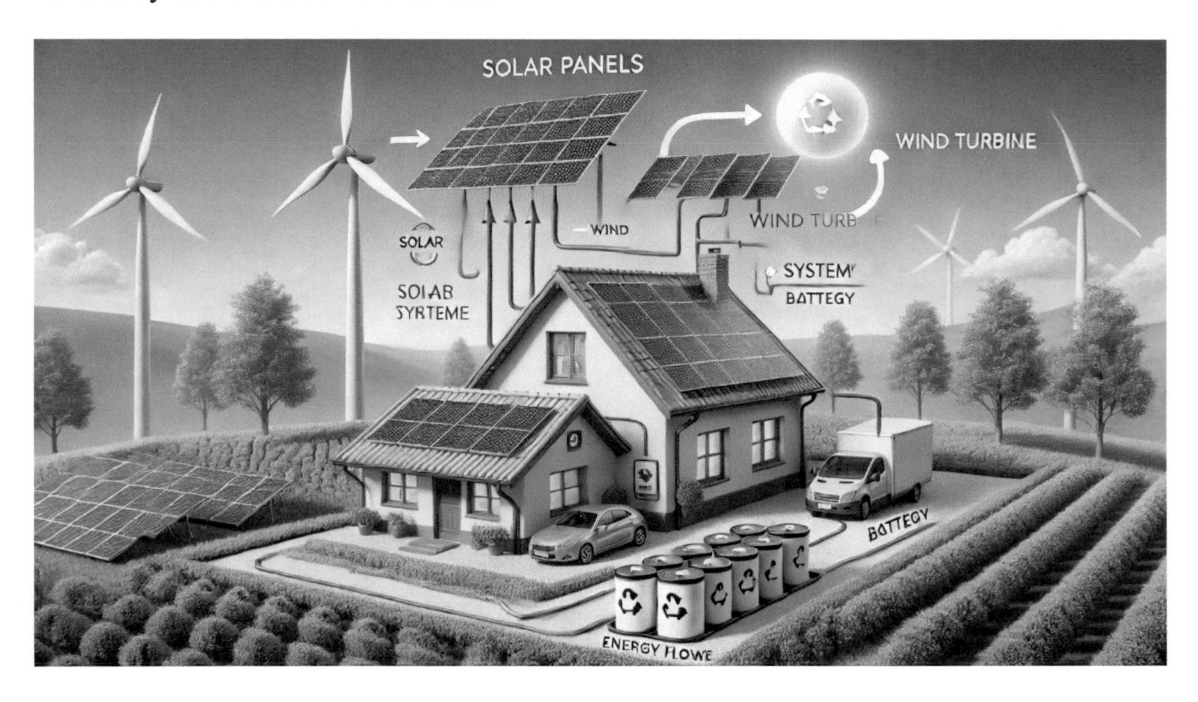

6.3.1 Maintenance and Troubleshooting

Like all off-grid energy systems, wind turbines require regular maintenance to ensure optimal performance and longevity.

Regular Maintenance Tasks:

- **Inspect Blades**: Check for signs of wear, cracks, or damage to the blades. Clean any debris or ice buildup that may hinder performance.
- **Lubricate Moving Parts**: Regularly lubricate bearings, rotors, and other moving parts to prevent wear and tear.
- **Check Electrical Connections**: Ensure all electrical connections are secure and free from corrosion.
- **Monitor Battery Bank**: Keep track of the battery charge levels and ensure they are functioning properly.

Key Takeaways:

- **Assess your location's wind potential** before investing in a wind turbine. Average wind speeds of **7-10 mph** are needed for efficient operation.
- Choose between **horizontal-axis** and **vertical-axis** turbines based on your space, wind patterns, and energy needs.
- Set up your wind power system by assembling the **turbine**, installing the **tower**, and connecting the system to a **battery bank** and **inverter**.
- Consider a **hybrid system** that combines both **wind** and **solar** power to maximize energy production and increase system reliability.
- Perform regular **maintenance** to ensure the longevity of your wind turbine and the overall efficiency of your off-grid energy system.

By integrating wind power into your off-grid energy system, you can take advantage of a renewable and sustainable energy source that works day and night, helping you achieve greater energy independence and resilience.

7. EMP Protection and Power System Redundancy

Electromagnetic pulses (EMPs) pose a significant threat to modern electrical systems, including off-grid power setups. An EMP can be caused by natural events, such as solar flares, or by man-made events like nuclear explosions at high altitudes. These pulses release a burst of electromagnetic radiation that can damage or destroy electronic equipment, especially systems that rely on electrical power, including your solar panels, wind turbines, and energy storage systems.

In this section, we'll explore the risks posed by EMPs, how to protect your off-grid power system, and the steps you can take to safeguard essential electronics and power generation systems from the devastating effects of an EMP.

7.1 Understanding EMP Risks to Off-Grid Systems

An **Electromagnetic Pulse (EMP)** is a burst of electromagnetic radiation that can severely disrupt or damage electrical and electronic devices. EMPs are categorized into three types:

1. **E1 Pulse**: This is a short, high-intensity burst of electromagnetic energy that occurs within nanoseconds of the initial pulse. It primarily affects devices like computers, radios, and power systems by overloading their circuits.
2. **E2 Pulse**: This is a mid-level pulse that lasts slightly longer and is less intense than the E1 pulse. It has effects similar to a lightning strike but can be neutralized with lightning protection systems.
3. **E3 Pulse**: This is a slower, longer-lasting pulse that can damage large electrical infrastructure, such as power grids, transformers, and long electrical lines.

An EMP event can cause anything from minor malfunctions to complete destruction of electrical systems, depending on the strength and proximity of the pulse.

Risks of EMPs for Off-Grid Systems

Even off-grid systems are not immune to the effects of an EMP. Any electrical or electronic component in your system can be damaged or destroyed by an EMP if it is not properly shielded.

Vulnerable Components in Off-Grid Power Systems:

- **Solar Charge Controllers**: The sensitive electronics inside charge controllers are particularly vulnerable to EMPs. They could be damaged or rendered inoperable.
- **Inverters**: Inverters, which convert DC to AC power, contain sensitive circuits that could be overloaded by an EMP, causing a complete failure.
- **Batteries**: While most batteries are not directly affected by EMPs, the electronic systems that manage them (Battery Management Systems, or BMS) could be damaged, rendering the battery unusable.

- **Solar Panels and Wind Turbines**: While the physical solar panels and wind turbines are unlikely to be damaged, the electrical components connected to them, such as wiring and electronic controls, can be affected.

7.2 Strategies to Safeguard Power Systems

There are several methods to protect your off-grid power system from the damaging effects of an EMP. These strategies focus on shielding, isolating, and protecting critical components.

1. Faraday Cages

A **Faraday Cage** is an enclosure made of conductive material (such as metal) that blocks electromagnetic radiation, effectively protecting electronic devices from an EMP. You can build Faraday cages to protect smaller electronic devices, such as radios, computers, or even inverters.

- **How It Works**: A Faraday cage works by distributing the electromagnetic energy of an EMP across its surface, preventing the pulse from reaching the devices inside.
- **DIY Faraday Cage**: You can create a simple Faraday cage using a metal container, such as a galvanized steel trash can, lined with non-conductive material like cardboard or foam. Ensure that the lid is tightly sealed to block the EMP completely.
- **Uses**: Store backup charge controllers, inverters, radios, and other critical electronic devices in Faraday cages to protect them from EMPs.

2. EMP-Proof Enclosures for Critical Electronics

For more sensitive components like solar charge controllers, inverters, or battery management systems, EMP-proof enclosures can provide an additional layer of protection. These enclosures are designed to shield critical electronics from high-intensity electromagnetic radiation.

- **Commercial EMP-Shielded Enclosures**: There are commercially available enclosures specifically designed to shield electrical components from EMPs. These are often made from materials like steel or copper and feature grounded designs to protect internal electronics.

3. Grounding Your System

Proper grounding of your off-grid power system is essential for protecting it from EMPs. Grounding ensures that excess electromagnetic energy is safely redirected into the earth, minimizing the impact on your electronic systems.

- **Install a Grounding System**: Make sure your solar panels, wind turbines, and electrical systems are properly grounded using copper grounding rods. Grounding can also protect your system from lightning strikes, which are similar to E2-type EMP pulses.
- **Surge Protectors**: Install EMP-rated surge protectors on key components, such as the inverter and charge controller, to protect against voltage spikes that can occur during an EMP.

Building Redundancy into Your System

Another strategy for protecting your off-grid power system is to build redundancy into your setup. Having multiple backup components can ensure that you are still able to generate and store power even if part of your system is affected by an EMP.

1. Backup Charge Controllers and Inverters

- **Store Extras in Faraday Cages**: Keep a backup charge controller and inverter stored in a Faraday cage or EMP-shielded enclosure. If the primary controller or inverter is damaged by an EMP, you can replace it with your backup.

2. Alternative Power Sources

- **Hand-Crank Generators**: Consider storing a hand-crank generator or solar-powered battery charger in a Faraday cage. These simple devices can provide essential power for radios, flashlights, or small electronic devices in the event of an EMP.
- **Manual Water Pumps and Cooking Systems**: Relying solely on electricity for water pumps and cooking can be risky in the event of an EMP. Invest in manual alternatives, such as hand pumps for water and rocket stoves for cooking, to ensure you can maintain basic functions if your power system is compromised.

2.4.5 Testing and Maintaining EMP Protection

It's not enough to install protective measures—you also need to regularly test and maintain them to ensure they are functioning correctly. This includes checking your Faraday cages, grounding systems, and surge protectors.

1. Test Faraday Cages

- **Simple Test Method**: You can test your Faraday cage by placing a small electronic device, such as a radio or cell phone, inside it. If the device no longer receives a signal while inside the cage, it indicates that the cage is effectively blocking electromagnetic waves.

2. Check Grounding Systems

- **Annual Inspection**: Regularly inspect your grounding system to ensure that connections are secure and that there is no corrosion on the grounding rods. Replace or repair any damaged components.

3. Surge Protector Maintenance

- **Replace Surge Protectors**: Surge protectors can degrade over time or after absorbing a large surge of energy. Make sure to replace surge protectors periodically, especially after a major storm or electrical event.

Key Takeaways:

- **EMPs** can damage or destroy off-grid power systems by overloading electrical components such as inverters, charge controllers, and battery management systems.
- Protect your system with **Faraday cages**, **EMP-proof enclosures**, and proper **grounding** to minimize damage in the event of an EMP.
- Build **redundancy** into your system by storing backup components, such as charge controllers and inverters, in protective enclosures.
- Regularly **test** and **maintain** your EMP protection measures to ensure they are effective and ready when needed.

By implementing these protective strategies, you can safeguard your off-grid power system from the potential devastation of an EMP, ensuring that your energy independence remains intact even in extreme scenarios.

8. Essentials of Off-Grid Cooking

8.1 Introduction to Off-Grid Cooking

Off-grid cooking is an essential skill for anyone living independently from modern utilities or preparing for potential emergencies where traditional energy sources might not be available. Whether you're living in a remote homestead, camping, or facing a natural disaster, learning how to cook without electricity or gas ensures that you can still prepare nutritious meals using alternative, sustainable methods. This chapter introduces the core principles of off-grid cooking and explores a variety of techniques that rely on natural resources such as wood, solar energy, or simple manual tools.

The beauty of off-grid cooking lies in its simplicity and resourcefulness. With a little creativity and the right equipment, you can cook a wide range of meals while minimizing your environmental impact and reliance on modern infrastructure. Additionally, off-grid cooking encourages a more mindful approach to food preparation, often requiring slower, more deliberate cooking processes that enhance the flavor and nutritional value of your food.

8.1 Core Principles and Equipment

The foundation of off-grid cooking revolves around the following principles, which emphasize sustainability, efficiency, and self-reliance:

1. Maximizing Resource Efficiency

Off-grid cooking methods rely on alternative fuel sources like wood, charcoal, and solar energy. Learning how to maximize the efficiency of these resources is essential for reducing fuel consumption and extending the life of your cooking system. Techniques like rocket stoves and solar ovens are designed to use minimal fuel while providing consistent heat.

2. Flexibility in Cooking Methods

Unlike conventional cooking, which often relies on ovens and stovetops, off-grid cooking requires flexibility. This includes mastering a range of techniques such as open-fire cooking, using portable stoves, and harnessing solar power. Each method offers unique advantages depending on your environment and available resources.

3. Portability and Durability

Off-grid cooking setups often need to be portable, especially for individuals who may need to move their living arrangements frequently. From campfire grills to lightweight solar ovens, the equipment used should be easy to transport and durable enough to withstand the challenges of off-grid living.

4. Energy Independence

One of the key goals of off-grid cooking is energy independence. By using renewable resources like sunlight or readily available materials like wood and biomass, you eliminate your reliance on external energy sources like electricity or gas. This is especially important in emergency situations where these utilities may not be available.

Essential Off-Grid Cooking Tools and Equipment

While off-grid cooking can be done with minimal tools, having the right equipment can make the process more efficient and enjoyable. Here are some must-have items for anyone looking to cook off the grid:

- **Cast-Iron Cookware**: Cast-iron skillets and Dutch ovens are essential for cooking over open flames or on stoves. They are durable, distribute heat evenly, and can withstand high temperatures, making them ideal for both campfire and wood-stove cooking.
- **Rocket Stove**: A rocket stove is a compact, highly efficient wood-burning stove that uses small amounts of fuel to generate intense heat. This makes it perfect for off-grid situations where conserving resources is critical.
- **Solar Oven**: Solar ovens harness the power of the sun to cook food slowly without using any fuel. They are perfect for sunny days and provide a completely sustainable way to prepare meals.
- **Campfire Grill or Tripod**: For campfire cooking, a portable grill or tripod is an excellent investment. It allows you to place pots, pans, or food directly over the fire, making it easy to grill, boil, or roast meals.
- **Thermal Cooker**: A thermal cooker is a slow-cooking device that uses retained heat to cook food over several hours without needing continuous energy input. Once food is brought to a boil, the pot is placed in an insulated container, where it continues cooking using the stored heat.
- **Fire Starters**: In off-grid environments, being able to start a fire quickly and efficiently is crucial. Fire starters like waterproof matches, lighters, or flint and steel should always be part of your off-grid kitchen supplies.

8.2 Choosing Fuel Sources and Managing Off-Grid Cooking Challenges

Cooking off the grid requires a keen understanding of the various fuel sources you can use to prepare food. Depending on your location and resources, different types of fuel may be more or less accessible. The key is to choose fuels that are both efficient and sustainable.

1. Wood

Wood is the most common fuel for off-grid cooking, particularly for campfires, rocket stoves, and wood-burning ovens. It is readily available in most environments and can be harvested sustainably if managed properly. Hardwoods like oak, maple, and hickory burn longer and provide more consistent heat than softwoods.

2. Biomass

Biomass, such as dried leaves, twigs, and other organic matter, can be used in certain stoves and ovens. This is a renewable resource that is often easy to find in forested or rural areas. Biomass fuels work well in rocket stoves and similar efficient cooking devices.

3. Charcoal

Charcoal is another efficient fuel source, especially for grilling or using in specific types of camp stoves. While it may require initial preparation or purchase, charcoal burns hotter and cleaner than wood, making it ideal for quick, high-heat cooking.

4. Solar Energy

For the most sustainable and eco-friendly option, solar energy is an excellent choice. Solar ovens and cookers use no fuel and can cook food slowly over several hours. They are ideal for sunny climates and provide a truly off-grid cooking experience.

Advantages and Challenges of Off-Grid Cooking

Advantages:

- **Sustainability**: Off-grid cooking methods rely on renewable resources, reducing your environmental footprint and promoting a more sustainable way of living.
- **Self-Sufficiency**: Learning how to cook without relying on modern utilities empowers you to be more independent and resilient, particularly in emergency situations or remote living.
- **Flavor and Nutrition**: Many off-grid cooking methods, such as slow-cooking with a thermal cooker or smoking meat, enhance the flavor and nutritional value of food through slower, more natural cooking processes.

Challenges:

- **Weather Dependency**: Certain off-grid cooking methods, particularly solar ovens, are dependent on weather conditions. Overcast or rainy days may require alternative cooking solutions.
- **Learning Curve**: Off-grid cooking requires learning new techniques and adapting to different heat sources. It may take time to master these methods, particularly for those accustomed to modern kitchen appliances.
- **Time and Effort**: Off-grid cooking often takes longer than traditional methods. Building and maintaining a campfire, for instance, can be more labor-intensive than simply turning on a stove.

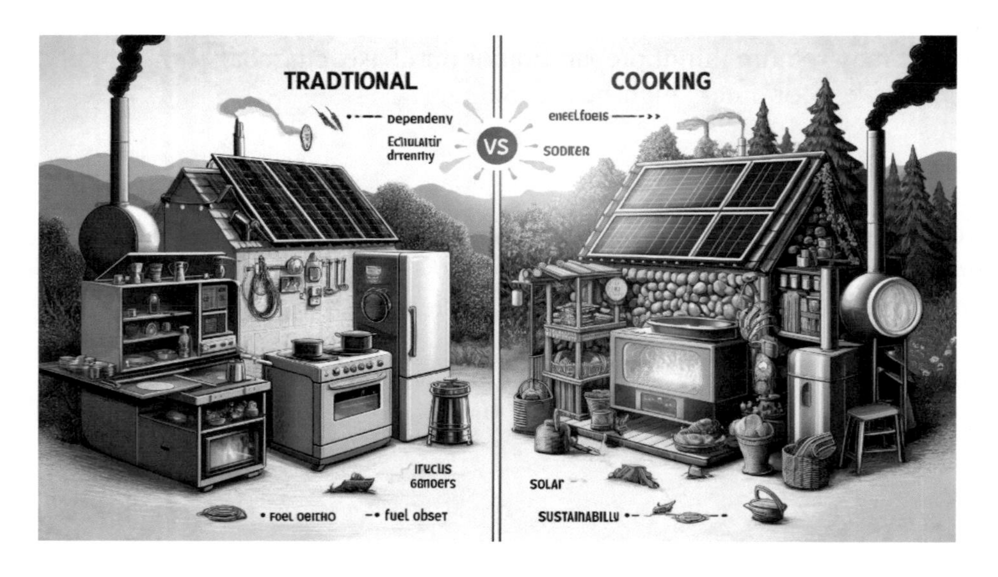

Preparing for Off-Grid Cooking: Skills to Master

Off-grid cooking is not just about having the right tools and fuel; it also requires developing a range of practical skills. Here are some key skills to master for successful off-grid cooking:

- **Fire-Building**: Knowing how to build, maintain, and control a fire is essential for most off-grid cooking methods. Practice building different types of fires (teepee, log cabin, etc.) to suit various cooking needs.

- **Temperature Control**: Whether you're cooking over a campfire or using a rocket stove, learning to control heat levels is crucial. Practice adjusting the intensity of your fire or stove to avoid overcooking or burning food.
- **Meal Planning**: Off-grid cooking often requires more time and effort, so planning meals ahead of time can help streamline the process. Prepare ingredients in advance and select recipes that suit the available cooking methods and fuel sources.
- **Food Preservation**: In off-grid settings, preserving food is an important skill. Techniques like smoking, drying, and fermenting can help extend the shelf life of your food, reducing the need for refrigeration.

Conclusion

Off-grid cooking is a blend of traditional skills and modern ingenuity, allowing you to prepare food in a sustainable, self-reliant way. By mastering these techniques and principles, you'll be able to cook delicious, nourishing meals regardless of your access to modern conveniences. Whether you're facing an emergency, living in a remote area, or simply wanting to reduce your reliance on external energy sources, off-grid cooking empowers you to take control of your food and energy needs.

9: Food Gathering and Preservation Techniques

One of the essential aspects of off-grid living and survival is ensuring a stable, reliable food supply. Gathering and preserving food off the grid requires knowledge of foraging, hunting, fishing, and various preservation methods to ensure that you can maintain a consistent supply of food, even during challenging conditions. This chapter will cover the foundational skills needed to gather food from nature, preserve it effectively, and maintain food safety when living without modern conveniences like refrigeration or supermarkets.

Food safety is a critical aspect of off-grid living, where the lack of modern conveniences like refrigeration, electricity, or running water can create challenges in preserving and preparing food. When living off the grid, the way you handle, store, and prepare food becomes even more important to prevent spoilage and foodborne illnesses. Understanding the risks and implementing proper food safety practices is key to maintaining a healthy and sustainable food supply in any off-grid environment.

In this section, we will explore the fundamental principles of off-grid food safety, providing guidelines for cleanliness, storage, preservation, and meal preparation. By following these practices, you can ensure that your food remains safe and edible even when you're far from modern infrastructure.

9.1 Basic Food Safety in Off-Grid Settings

Without access to modern kitchen appliances and refrigeration, you need to adopt different techniques for cleaning and storing food. These methods help prevent contamination and ensure that your food supply remains fresh and safe for consumption.

1. Cleanliness

Maintaining cleanliness is one of the most important factors in preventing the growth of harmful bacteria and contamination in your food. Whether you're handling raw meat, freshly foraged plants, or harvested produce, cleanliness is critical in off-grid food preparation.

- **Wash Hands and Surfaces**: Always wash your hands with clean, potable water and soap before handling food. Clean all cooking surfaces, utensils, and cutting boards with soap and water regularly to prevent cross-contamination.
- **Filtered or Boiled Water**: Since off-grid water sources may not be as clean as municipal supplies, use filtered, boiled, or chemically treated water for washing food and cooking.
- **Raw Meat Handling**: Take extra precautions when handling raw meat. Use separate cutting boards for raw meats and vegetables to avoid cross-contamination. Clean all surfaces that come into contact with raw meat thoroughly.

2. Food Separation

Keeping different types of food separate is crucial to prevent cross-contamination, particularly when dealing with raw meats, vegetables, and ready-to-eat foods. This is especially important in off-grid settings where storage space is limited.

- **Store Raw Meat Separately**: Keep raw meat, fish, and poultry separate from fruits, vegetables, and cooked foods. Use sealed containers or tightly wrap the raw meat to prevent juices from leaking onto other foods.
- **Organize Your Storage**: If you don't have access to refrigeration, store foods in cool, dark places. Use root cellars, underground storage, or shaded areas to keep fresh produce from spoiling too quickly.

3. Storage Temperature

Temperature control is a critical element of food safety, especially when refrigeration isn't available. In off-grid situations, it's important to understand which foods require immediate consumption and which can be preserved for later use.

- **Root Cellaring**: A traditional method used for centuries, root cellaring involves storing vegetables, fruits, and some meats in a cool, underground environment. A root cellar maintains a stable temperature and humidity level, which slows the spoilage process.
- **Preserving Foods Quickly**: Perishable foods like meat and dairy should be consumed quickly or preserved through methods such as smoking, drying, salting, or canning (which we will explore later in this chapter). Use available resources to preserve these foods as soon as possible to prevent waste.

Understanding Foodborne Illnesses in Off-Grid Settings

Off-grid living increases your exposure to certain risks associated with foodborne illnesses, primarily because of the lack of refrigeration and modern food storage methods. The bacteria, viruses, and parasites that cause foodborne illnesses can thrive in environments where food isn't properly stored or handled.

Common Foodborne Pathogens:

- **Salmonella**: Often found in raw or undercooked eggs, poultry, and meat. This bacterium can spread quickly in hot or unsanitary conditions.
- **E. coli**: Found in contaminated water, undercooked ground beef, and raw vegetables. This pathogen thrives when foods are not properly washed or cooked.
- **Listeria**: Can grow in improperly stored or unpasteurized dairy products, processed meats, and soft cheeses.

Best Practices for Food Preservation and Safety

Preserving food properly is essential to avoid food spoilage and waste, especially when you don't have access to modern refrigeration. There are various methods to preserve food for longer periods without refrigeration.

1. Smoking and Drying

- **Smoking**: Smoking is an ancient technique that both flavors and preserves meat. The smoke creates a protective barrier around the meat while drying it out, making it less susceptible to bacteria. Ensure that you maintain a low, steady heat (typically 150-160°F) and smoke the meat for several hours until it's fully preserved.
- **Drying**: Dehydrating food by removing moisture helps prevent bacterial growth. You can air-dry herbs, fruits, and vegetables in the sun, or use a solar dehydrator for faster results.

2. Salting and Brining

Salt has been used for centuries to preserve food by drawing moisture out of meats and vegetables, which prevents bacteria from growing.

- **Salting**: Covering meat with salt creates a hostile environment for bacteria, preserving the meat for extended periods. Salted meats can be stored in a cool, dry place or hung to air-cure.
- **Brining**: Soaking meat or vegetables in a saltwater solution (brine) preserves food while also adding flavor. Brining works particularly well for preserving meats like pork, fish, or poultry.

3. Canning and Fermenting

Canning and fermenting are excellent ways to preserve food for long-term storage, especially for vegetables, fruits, and even meats.

- **Canning**: Proper canning involves placing food in jars and heating them to high temperatures to kill bacteria. Use either a boiling water canner for high-acid foods (like fruits) or a pressure canner for low-acid foods (like meats and vegetables).
- **Fermenting**: Fermentation is a natural preservation method that uses beneficial bacteria to break down sugars in food, producing acids that prevent spoilage. Popular fermented foods include sauerkraut, pickles, and yogurt.

Water Safety in Off-Grid Food Preparation

Water is essential not only for drinking but also for food preparation and cooking. Ensuring that your water supply is clean and safe is critical to preventing contamination and illness. In off-grid environments, where water sources may be drawn from rivers, lakes, or wells, extra precautions must be taken to ensure water is potable.

1. Water Filtration and Purification

- **Boiling**: One of the simplest ways to purify water is by boiling it for at least one minute. This kills bacteria, viruses, and parasites, making it safe for consumption and food preparation.
- **Filtration Systems**: Portable water filters can remove harmful contaminants from water. Ensure you use a high-quality filter designed to eliminate both physical debris and harmful pathogens.

- **Chemical Treatment**: Use iodine or chlorine tablets to treat water when boiling or filtration isn't possible. Follow the instructions carefully to ensure safe water for cooking.

2. Proper Water Storage

In off-grid situations, having a reliable source of clean, stored water is essential. Store water in clean, food-grade containers in a cool, shaded area to prevent contamination. Rotate stored water regularly and treat it with purification methods before use if necessary.

Key Takeaways for Off-Grid Food Safety:

- **Maintaining cleanliness** and **separation** during food preparation is essential to prevent contamination, particularly when working with raw meats and vegetables.
- **Proper storage** of perishable items in cool, dark environments or through preservation methods like smoking, drying, and salting is critical in off-grid settings.
- **Preserving food** immediately after gathering or preparing ensures that your food supply remains safe for consumption over long periods.
- **Water safety** is a priority. Ensure all water used for cleaning, cooking, and drinking is properly filtered, boiled, or chemically treated to prevent contamination.

By adhering to these food safety principles, you can ensure that your off-grid lifestyle remains healthy and sustainable, with access to safe, nutritious food regardless of modern conveniences.

9.2 Hunting, Fishing, and Foraging for Wild Food

Hunting and fishing are essential skills for off-grid living, providing a reliable source of protein that complements your survival garden or foraging efforts. Whether you're hunting game animals or catching fish, mastering these techniques can ensure your food supply remains stable and nutritious throughout the year. In an off-grid environment, learning to hunt and fish not only builds self-reliance but also helps you develop a deeper connection with nature and its resources.

In this section, we will explore the key techniques, tools, and strategies for off-grid hunting and fishing, including ways to process and preserve your catch to maximize its value.

Off-Grid Fishing Techniques: Trotlines, Jug Lines, and Fish Transport

Fishing is a time-tested method for gathering food off the grid. With access to lakes, rivers, or streams, fishing can provide a steady supply of fish that can be eaten fresh or preserved for long-term storage. Off-grid fishing methods like trotlines, jug lines, and simple hand-line fishing require minimal equipment and can be set up to catch fish even when you're not actively fishing.

1. Trotlines

Trotlines are one of the most effective and low-maintenance ways to catch fish in off-grid settings. A trotline consists of a long fishing line suspended horizontally in the water, with multiple baited hooks attached at intervals.

- **Setting a Trotline**: Trotlines can be anchored to the shore or between two stationary points, such as trees or stakes driven into the ground. The baited hooks are placed at varying depths to target different species of fish.
- **Advantages**: Trotlines allow you to catch multiple fish over time without constant attention. They are particularly effective for catching catfish, carp, and other bottom feeders.
- **Legal Considerations**: Always check local regulations for the use of trotlines, as some areas may have restrictions or require permits.

2. Jug Lines

Jug lines are similar to trotlines but are designed to float on the surface of the water. They are made by attaching baited hooks to floating objects, such as plastic jugs, bottles, or buoys, which drift freely in the water.

- **Setting Jug Lines**: Each jug is tied to a length of line with one or more hooks. Jugs are then released into the water, allowing them to drift and attract fish. Once a fish is hooked, the jug will move or bob, signaling a catch.
- **Advantages**: Jug lines are highly portable and can cover a large area. They are particularly useful in lakes or ponds where fish move freely and can be tracked by following the jugs.

3. Limb Lines

Limb lines involve tying baited fishing lines to the branches of trees that hang over the water. This method works well in rivers and streams where fish naturally swim near the banks.

- **Advantages**: Limb lines are easy to set up and require minimal equipment. They work well for catching species that swim close to the shore, such as catfish or bass.
- **Monitoring**: Like trotlines and jug lines, limb lines should be checked regularly to ensure that fish are caught and removed promptly to prevent spoilage.

4. Transporting Fish

Once you have caught fish, it's crucial to clean, store, or transport them as quickly as possible to maintain freshness. Here are some tips for handling your catch:

- **Immediate Cleaning**: If you're not close to your homestead or a preservation facility, clean the fish immediately after catching them by removing the guts, gills, and scales. This helps prevent spoilage and makes the fish easier to transport.
- **Cool Storage**: Use coolers, ice, or nearby cold streams to keep fish fresh until they can be cooked or preserved. In colder climates, snow or ice can also be used for temporary storage.

Step-by-Step Guide to Deer Hunting: From Preparation to Butchering

Deer hunting is a critical skill for providing a substantial amount of meat for off-grid living. With careful planning and technique, hunting can yield enough meat to last for months, especially when combined with proper preservation methods. This section provides a comprehensive guide to deer hunting, including preparation, tracking, shooting, field dressing, and butchering.

1. Preparation

Before setting out to hunt deer, it's essential to prepare both physically and mentally. This involves gathering the right gear, understanding the environment, and ensuring that you comply with local hunting regulations.

- **Gear**: Ensure you have the appropriate firearm (rifle, shotgun, or bow), ammunition, hunting license, and safety gear. A pair of binoculars, a rangefinder, and a knife for field dressing are also essential tools.
- **Scouting**: Spend time scouting the area before hunting to identify deer trails, feeding grounds, and bedding areas. Understanding the habits of deer will increase your chances of success.
- **Regulations**: Always familiarize yourself with local hunting laws, including open season dates, bag limits, and required permits.

2. Tracking and Shooting

Tracking deer requires patience, observation, and an understanding of their behavior. Here are some key tracking tips:

- **Tracks and Signs**: Look for fresh tracks, droppings, and signs of deer feeding. Early morning and late afternoon are prime times for tracking, as deer are most active during these periods.
- **Wind Direction**: Always approach from downwind so that the deer cannot detect your scent. A deer's sense of smell is highly developed and can alert them to danger from great distances.
- **Shot Placement**: When you're ready to take a shot, aim for the heart or lungs for a quick, humane kill. This reduces suffering for the animal and ensures you can recover it quickly.

3. Field Dressing a Deer

Field dressing is the process of removing the internal organs from the deer to preserve the meat and make it easier to transport. It's important to field dress the deer as soon as possible after the kill to prevent spoilage.

- **Step-by-Step Field Dressing**:
    - Lay the deer on its back and make a shallow incision along the belly, from the ribcage to the pelvis.
    - Carefully remove the internal organs, being cautious not to puncture the stomach or intestines to avoid contaminating the meat.
    - Drain excess blood and remove the windpipe.
    - Rinse the body cavity with clean water if available, and transport the deer back to your base for further butchering.

4. Butchering the Deer

Once the deer has been field dressed and transported, the next step is butchering the carcass into usable cuts of meat. Butchering can be done on-site or at home, depending on your location and available resources.

- **Key Cuts**:
  - **Tenderloin and Backstrap**: These are the most tender cuts, located along the spine. They should be removed first and can be cooked fresh or preserved.
  - **Legs (Hindquarters)**: The hindquarters provide large amounts of meat and are ideal for steaks, roasts, or jerky.
  - **Shoulders and Ribs**: These cuts are best used for stews or ground meat.
  - **Neck and Trimmings**: These parts can be used for making sausage or ground meat.
- **Preservation**: After butchering, preserve the meat by freezing, smoking, or drying it to ensure it lasts for months. If freezing is not an option, use traditional methods like salting or curing to extend the shelf life.

Key Takeaways for Off-Grid Hunting and Fishing:

- **Fishing**: Trotlines, jug lines, and limb lines are simple and effective ways to catch fish without constant attention. Always check local regulations and ensure you store or process fish quickly to maintain freshness.
- **Hunting**: Deer hunting provides a large supply of meat, but it requires careful preparation, tracking, and butchering skills. Field dressing and proper preservation techniques are critical for ensuring the meat lasts over time.
- **Self-Reliance**: Hunting and fishing are valuable skills for off-grid living, offering a sustainable food source that can complement gardening and foraging efforts.

By mastering these techniques, you can ensure a steady and nutritious supply of protein for your off-grid lifestyle, contributing to long-term sustainability and food security.

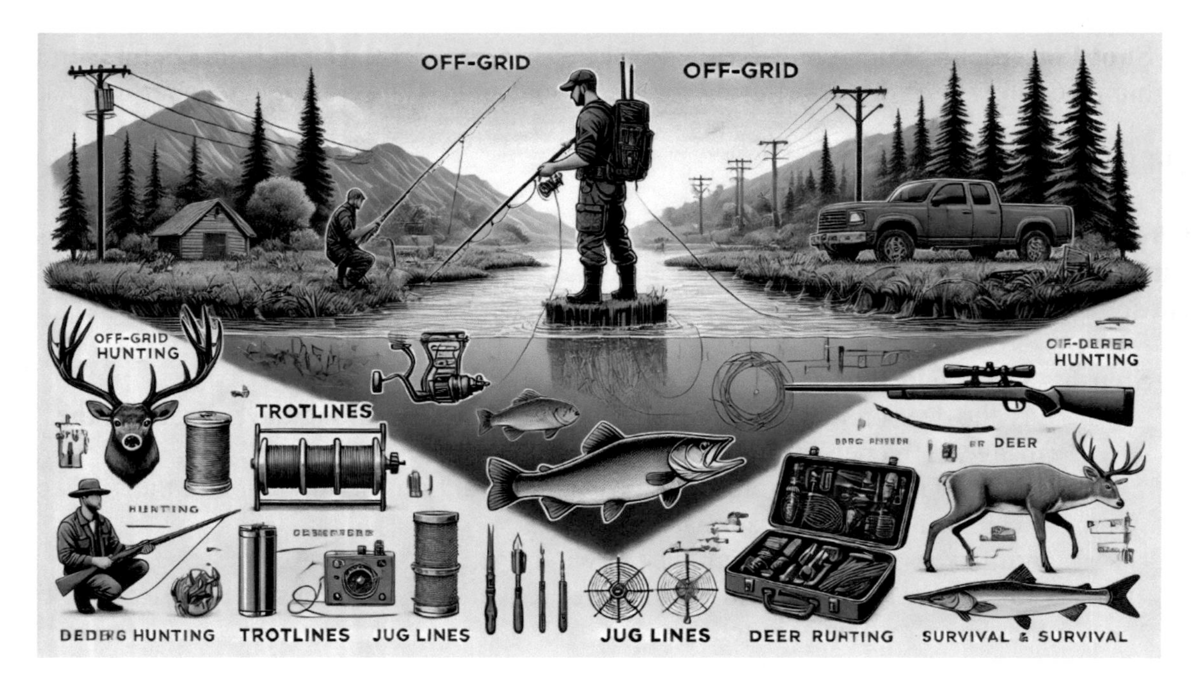

9.3 Foraging Wild Plants and Small Game

Foraging and hunting small game are two of the most practical ways to supplement your off-grid diet, providing essential vitamins, minerals, and protein when other food sources are limited or difficult to access. By understanding how to identify edible plants and efficiently trap small game, you can ensure that you always have access to nutritious food, even in the most remote settings.

This section explores techniques for foraging edible wild plants, trapping small game such as rabbits and squirrels, and the methods for preparing your harvest to ensure it's safe to eat.

Tips for Foraging Edible Wild Plants

Foraging is a valuable skill that allows you to harness the natural abundance of your environment. Wild plants can provide essential nutrients, but it's crucial to have a good understanding of which plants are safe to eat and which ones should be avoided.

1. Identifying Edible Plants

Foraging requires knowledge and experience in plant identification to avoid poisonous or harmful species. Always carry a reliable field guide or plant identification app when foraging, especially if you are unfamiliar with local flora.

- **Common Edible Wild Plants**:
  - **Dandelions**: Nearly every part of this plant is edible, including the leaves, roots, and flowers. Dandelions are rich in vitamins A, C, and K.
  - **Wild Onions and Garlic**: These can be found in forested or meadow areas and can be easily identified by their distinctive smell.
  - **Cattails**: Found in wetland areas, cattail roots and shoots are edible and provide a good source of carbohydrates.
  - **Nettle**: When properly prepared (usually through boiling), nettle is a nutritious green that is rich in vitamins and minerals.
  - **Wild Berries**: Berries such as blackberries, raspberries, and blueberries are easy to identify and offer essential nutrients and antioxidants.
- **Safety Considerations**:
  - **Avoid Unknown Plants**: Never consume plants unless you are 100% sure of their identity. Some poisonous plants closely resemble edible ones.
  - **Testing for Edibility**: In situations where you are uncertain about a plant's edibility, the **Universal Edibility Test** can be used. This involves testing a small portion of the plant on your skin and then your lips before consuming it. Always proceed cautiously.

2. Seasonal Foraging

Foraging success depends largely on the time of year and your geographical location. Being aware of seasonal cycles will help you locate the best food sources.

- **Spring**: Early spring is ideal for foraging young, tender greens such as dandelion leaves, chickweed, and wild onions.

- **Summer**: Summer is abundant with fruits like berries and edible flowers.
- **Fall**: Fall is harvest season for nuts, seeds, and root vegetables like wild carrots and burdock.
- **Winter**: In colder months, foraging options are limited, but evergreen plants, certain roots, and stored nuts can still be found.

3. Sustainability in Foraging

Foraging should always be done sustainably to preserve the environment and ensure the availability of resources for future harvests.

- **Harvest Only What You Need**: Avoid over-harvesting, especially when gathering rare or slow-growing plants. Take no more than 30% of a plant population to allow for regrowth.
- **Leave Roots Intact**: When harvesting leafy plants, avoid pulling them up by the roots. Instead, trim leaves or stems to allow the plant to continue growing.
- **Avoid Polluted Areas**: Never forage near roads, industrial sites, or areas that may be contaminated with pesticides or pollutants.

Preparing Rabbits, Squirrels, and Game Birds

Small game, such as rabbits, squirrels, and game birds, are valuable sources of protein for off-grid living. These animals are often more abundant and easier to catch than large game, making them an ideal food source for those living in remote or off-grid environments. Trapping small game is a practical way to supplement your diet, and learning how to prepare and preserve the meat ensures that none of your efforts go to waste.

1. Trapping Small Game

Setting traps for small game is a low-effort way to catch food, allowing you to continue with other tasks while waiting for your trap to capture an animal. There are several types of traps that can be used depending on the environment and species you're targeting.

- **Snares**: Snares are simple, passive traps made from wire or strong cord. When placed in high-traffic areas where animals naturally travel (such as along game trails), snares can efficiently catch rabbits, squirrels, or other small mammals.
- **Box Traps**: Box traps are humane traps that capture animals alive, allowing you to release non-target species. They can be made from wood, metal, or wire and are baited to attract animals.
- **Pitfall Traps**: Pitfall traps are holes dug into the ground and covered with branches or leaves. Animals fall into the trap and are unable to escape, making them easy to retrieve later.

Trap Placement:

- Locate traps along animal trails, near water sources, or in areas where animals are known to forage.
- Check your traps frequently to ensure humane treatment of the captured animals and to prevent spoilage.

2. Cleaning and Preparing Small Game

Once you've caught a rabbit, squirrel, or game bird, you'll need to clean and process the animal to make it ready for cooking or preservation.

- **Cleaning Small Game**:
  - **Rabbits and Squirrels**: Begin by removing the fur. This can be done by making a small incision in the skin and peeling it back. After skinning the animal, remove the internal organs carefully to avoid puncturing the stomach or intestines, which can contaminate the meat. Rinse the body cavity with clean water.
  - **Game Birds**: Pluck the feathers carefully or skin the bird entirely, depending on your preference. After plucking, make a small incision near the vent and remove the internal organs. Rinse thoroughly.
- **Preserving Small Game Meat**:
  - **Freezing**: If you have access to a freezer, clean and package the meat in sealed bags or containers for long-term storage.
  - **Smoking or Drying**: Smoking and drying are ideal off-grid preservation methods. Smoke the meat over a low heat for several hours or use a solar dehydrator to remove moisture, which helps prevent spoilage.
  - **Salting**: Coat the meat in a layer of salt and allow it to air-cure. This process draws out moisture and preserves the meat for longer periods.

3. Cooking Small Game

- **Stewing**: One of the easiest ways to cook small game is by stewing. Rabbit and squirrel meat can be tough, but slow-cooking in a stew tenderizes the meat while retaining its flavor.
- **Roasting**: Small game birds such as quail or grouse can be roasted over an open fire or in a solar oven. For best results, baste the meat with oil or fat to prevent drying out.
- **Jerky**: Thin strips of meat can be dried or smoked to make jerky, which is lightweight, easy to store, and ideal for long-term food storage.

Key Takeaways for Foraging and Hunting Small Game:

- **Foraging**: Learn to identify edible wild plants and forage sustainably. Always use caution when consuming wild plants and cross-check with reliable identification resources.
- **Trapping Small Game**: Snares, box traps, and pitfall traps are effective, low-maintenance methods for catching small game like rabbits and squirrels.
- **Cleaning and Cooking**: Properly clean and prepare small game to avoid contamination. Use methods such as stewing, smoking, or making jerky to enjoy nutritious meals or preserve meat for future use.

Foraging and hunting small game provide essential skills for off-grid living. By combining these techniques with food preservation methods, you can ensure a steady and reliable food supply in any off-grid environment, helping you maintain a healthy, balanced diet.

10. Livestock Raising and Beekeeping

Raising livestock and keeping bees are two of the most rewarding and sustainable ways to ensure a steady supply of food and resources when living off the grid. Not only do they provide essential products like meat, milk, eggs, and honey, but they also contribute to self-sufficiency, enhancing your ability to live independently from external sources. This chapter explores the basics of off-grid livestock management, including choosing the right animals, providing adequate care, and the fundamentals of beekeeping for honey production.

10.1 Introduction to Off-Grid Livestock Raising

Livestock raising is an integral part of off-grid living for those who wish to be self-sufficient in food production. Whether you're raising chickens for eggs, cattle for meat and dairy, or pigs for pork, livestock can provide a consistent and renewable food source. This section covers the basics of choosing the right livestock, caring for them without modern infrastructure, and ensuring their health and well-being.

How to Choose the Right Livestock for Self-Sufficiency

When selecting livestock for off-grid living, several factors must be considered to ensure that you choose animals that are well-suited to your environment, resources, and needs.

1. Size of Your Property and Resources

- **Land Availability**: Larger animals, such as cattle and pigs, require more land and resources than smaller animals like chickens or goats. Ensure that your property can support the space and grazing needs of the animals you intend to raise.
- **Feed and Water**: Consider the amount of feed and water each species requires. Animals like goats and chickens may need less water and forage than cows or pigs. In off-grid settings, where water and feed may be limited, smaller animals are often more manageable.

2. Purpose of the Livestock

- **Egg and Meat Production**: Chickens are ideal for providing both eggs and meat, making them a versatile choice for small homesteads. Dual-purpose breeds like the Rhode Island Red are particularly well-suited for off-grid living.
- **Dairy Production**: Goats and cows are common choices for milk production. Goats are easier to care for on smaller plots of land, while cows provide a higher yield of milk but require more resources.
- **Meat Production**: Pigs and cattle are excellent choices for meat, with pigs being easier to raise in smaller spaces. Pigs also provide a valuable source of lard, which can be used for cooking and soap-making.

3. Hardiness and Adaptability

Off-grid environments often come with unpredictable weather conditions and limited access to modern veterinary care. It's essential to choose livestock breeds known for their hardiness and ability to adapt to outdoor living and fluctuating climates.

- **Goats**: Hardy and versatile, goats thrive in a wide range of environments. They are excellent foragers, making them ideal for off-grid homesteads with limited resources.
- **Heritage Breeds**: Many heritage livestock breeds, such as Highland cattle or Berkshire pigs, are naturally more resistant to diseases and better adapted to grazing in rougher environments.

A Guide to Raising Chickens, Pigs, and Cattle

Each type of livestock has specific needs and care requirements. Understanding these needs is key to ensuring that your animals remain healthy and productive.

1. Raising Chickens

Chickens are one of the easiest and most rewarding animals to raise off the grid. They provide a reliable source of eggs, meat, and even fertilizer through their manure.

- **Shelter**: Chickens need a well-ventilated but secure coop to protect them from predators and harsh weather. Ensure the coop has enough space for each bird to roost and lay eggs.
- **Feeding**: Chickens are natural foragers and can feed on insects, grass, and food scraps. Supplement their diet with grains like corn and wheat to ensure they receive proper nutrition.
- **Egg Production**: Hens will naturally lay eggs, but providing comfortable nesting boxes and ensuring their feed is nutrient-rich can help increase egg production.
- **Predator Protection**: Chickens are vulnerable to predators like foxes, hawks, and raccoons. Reinforce their coop and run with strong fencing and close them in at night.

2. Raising Pigs

Pigs are versatile and efficient animals for off-grid meat production. They can be raised on pasture or in a contained area, and they are excellent at converting food waste into valuable meat.

- **Shelter**: Pigs need a sturdy shelter that protects them from wind and rain. A simple shed or hog pen is sufficient, as pigs are hardy animals.
- **Feeding**: Pigs can be fed kitchen scraps, garden waste, and forage. Supplementing their diet with grains will help them grow faster and produce better-quality meat.
- **Breeding**: If you intend to breed pigs, ensure you have the proper space and resources to manage piglets. Pigs usually give birth to large litters, and they need adequate space for raising young.
- **Slaughtering and Butchering**: Pigs provide a large amount of meat, and butchering them can be done on-site. The meat can be preserved through smoking, salting, or curing.

3. Raising Cattle

Cattle are more resource-intensive to raise than chickens or pigs, but they provide a valuable source of milk and beef. They are best suited for off-grid homesteads with plenty of land for grazing.

- **Shelter**: Cattle need open pasture to graze and a simple barn or lean-to for shelter during bad weather. They are hardy animals, but protection from wind and rain is necessary.
- **Feeding**: Cows are natural grazers, but during winter months or periods of low forage, you'll need to supplement their diet with hay or silage.
- **Milking**: If raising dairy cows, daily milking is essential to keep the cow healthy and ensure a steady supply of milk. Learn proper milking techniques to avoid injury to the cow and ensure hygiene.
- **Calving**: Breeding cows requires attention to calving. A calf born in good health will eventually contribute to your meat or dairy supply.

Livestock Management for Long-Term Self-Sufficiency

Raising livestock off-grid is a critical component of long-term self-sufficiency. Proper management ensures that your animals not only survive but thrive, providing you with a reliable source of food, labor, and resources for years to come. This section will cover advanced livestock management strategies, including rotational grazing, selective breeding, and producing your own feed, to help you create a sustainable and resilient livestock operation.

---

1. Rotational Grazing Systems: Enhancing Pasture Health

Rotational grazing is a highly effective method for managing grazing animals while improving the health and productivity of your pasture. Instead of allowing animals to graze continuously in one area, you move them between different sections of the pasture, allowing grazed areas time to recover and regrow.

- **How It Works**: Divide your grazing area into several smaller paddocks or sections. Animals graze in one section for a set period (typically a few days to a week), after which they are moved to the next section. The grazed section is left to rest and regrow before animals return.
- **Benefits of Rotational Grazing**:
  - **Improved Soil Health**: Moving animals regularly prevents overgrazing, which can degrade the soil. Instead, the soil is given time to regenerate, increasing nutrient levels and promoting healthy plant growth.
  - **Increased Forage Production**: By allowing each section of the pasture to rest, grasses and other plants can regrow, leading to higher forage production over time.
  - **Better Animal Health**: Rotational grazing reduces the buildup of parasites in the soil, which helps keep livestock healthier. It also provides animals with a constant supply of fresh, nutritious forage.
- **Implementing a Rotational Grazing System**:
  - **Fencing**: Use temporary electric fencing to create paddocks, allowing you to easily move livestock between grazing sections.
  - **Water Access**: Ensure each paddock has access to water, either through natural sources (like streams or ponds) or by moving water troughs between sections.

    - **Grazing Schedule**: Create a grazing schedule based on the size of your paddocks, the number of animals, and the growth rate of your pasture. Adjust this schedule seasonally to account for slower grass growth in winter or dry periods.

---

2. Selective Breeding: Improving Productivity and Disease Resistance

Selective breeding is an important tool for improving the productivity and resilience of your livestock. By choosing animals with desirable traits—such as disease resistance, high fertility, or rapid growth rates—you can build a herd or flock that is better suited to your specific environment and needs.

- **Traits to Consider**:
    - **Disease Resistance**: Select animals that have shown resistance to common diseases or parasites in your area. This reduces the need for medical interventions and helps ensure a healthy herd.
    - **Fertility and Birth Rate**: Animals with higher fertility rates or those that produce larger litters (in the case of pigs or rabbits) will increase your livestock numbers more quickly.
    - **Growth Rate and Feed Efficiency**: Choose animals that grow quickly and convert feed efficiently into body mass. This is particularly important for meat production, where faster growth results in a shorter time to market or consumption.
    - **Temperament**: For animals like cattle, sheep, and goats, calm and manageable temperaments make handling easier and reduce stress on both the animals and the farmer.
- **Maintaining Genetic Diversity**: While selective breeding focuses on specific traits, it's important to maintain genetic diversity within your herd to avoid inbreeding. Introduce new breeding stock from time to time to prevent genetic problems and keep your herd healthy.
- **Record Keeping for Breeding**: Keep detailed records of your breeding stock, tracking traits like birth weight, growth rate, health issues, and reproductive success. This helps you make informed decisions about which animals to breed in future seasons.

---

3. Producing Your Own Animal Feed

One of the challenges of off-grid livestock management is ensuring a consistent, sustainable source of feed for your animals, especially during the winter months or periods of drought. By producing your own feed, you reduce dependency on external sources and ensure your animals are well-fed year-round.

- **Crops for Animal Feed**: Certain crops are ideal for growing your own animal feed. Consider the following options:
    - **Alfalfa**: High in protein and an excellent feed for cattle, goats, and rabbits. Alfalfa can be grown as hay or used in its green form for grazing.
    - **Corn**: Corn is a common feed crop for pigs, chickens, and cattle. It's rich in carbohydrates and can be stored as whole kernels or ground into feed.
    - **Fodder Beets and Turnips**: These root vegetables grow well in cool climates and can be harvested during the winter months, providing a fresh feed source for livestock.

- **Barley and Oats**: These grains are great for poultry and pigs. They can be grown and stored easily and are highly nutritious.

- **Forage Preservation**:
  - **Haymaking**: Cutting and drying hay is a simple and effective way to store forage for winter. Be sure to harvest hay at the right time—before plants mature fully—to ensure maximum nutritional value.
  - **Silage**: Silage is fermented, high-moisture feed made from crops like corn or grass. It's ideal for cattle and sheep and can be stored for long periods in airtight silos or sealed plastic bags.
  - **Grain Storage**: Store grain in rodent-proof, moisture-controlled bins or silos to ensure a steady supply of feed throughout the year. Proper storage prevents spoilage and keeps feed fresh.

---

4. Animal Health and Disease Management

Maintaining the health of your livestock is crucial for long-term productivity and sustainability. By focusing on preventive care and natural treatments, you can minimize the need for costly medical interventions.

- **Natural Remedies for Common Ailments**:
  - **Herbal Treatments**: Herbs like garlic, thyme, and oregano have natural antibacterial and antifungal properties. These can be used in feed or applied topically to treat minor infections.
  - **Diatomaceous Earth**: This natural powder is effective for controlling internal parasites like worms in livestock. It can be mixed into feed to help prevent infestations.
  - **Apple Cider Vinegar**: Adding a small amount of apple cider vinegar to water troughs can help boost immunity and improve digestion in animals.
- **Vaccinations and Preventive Care**:
  - **Vaccinations**: Ensure your livestock are vaccinated against common diseases in your area. Speak with a local veterinarian to determine the right vaccination schedule for your herd or flock.
  - **Parasite Control**: Rotate pastures regularly to prevent the buildup of parasites in the soil. Consider using natural parasite control methods, such as adding herbs like wormwood to your animals' diet.
- **Routine Health Checks**: Regularly check your animals for signs of illness, such as changes in behavior, weight loss, or unusual discharge. Early detection of health issues is key to preventing the spread of disease and minimizing losses.

Key Takeaways for Livestock Management:

- **Rotational Grazing**: Implement a rotational grazing system to improve pasture health, increase forage production, and reduce parasite load in your livestock.
- **Selective Breeding**: Use selective breeding to enhance desirable traits like disease resistance, fertility, and growth rates, while maintaining genetic diversity in your herd.
- **Feed Production**: Grow your own feed crops, such as alfalfa, corn, and barley, to reduce dependency on store-bought feed and ensure your animals have enough food year-round.

- **Animal Health**: Focus on preventive care, including natural remedies, vaccinations, and routine health checks, to keep your livestock healthy and productive.

By following these advanced livestock management strategies, you'll build a sustainable and resilient livestock operation that supports your off-grid self-sufficiency goals for the long term.

10.2 Beekeeping for Sustainability

Beekeeping is a valuable addition to off-grid living. Bees produce honey, beeswax, and propolis, which can be used for food, medicinal purposes, and crafting. In addition, bees play a crucial role in pollinating plants, which improves the productivity of your garden.

Getting Started with Beekeeping to Produce Honey and Beeswax

Beekeeping is relatively low-maintenance, but there are important steps to take to ensure your bees thrive and produce honey efficiently.

1. Choosing the Right Hive

The type of hive you choose will influence how easy it is to manage your bees and harvest honey. The two most common hive types are:

- **Langstroth Hive**: This is the most popular hive for beekeepers. It consists of stacked boxes that allow for easy expansion as the colony grows.
- **Top Bar Hive**: A simpler design, the top bar hive requires less maintenance but may produce lower honey yields. It's a great option for those who prefer natural beekeeping methods.

2. Setting Up the Apiary

- **Location**: Place your hive in a quiet, sheltered location with access to flowering plants and water. Ensure the area is protected from strong winds and receives some shade during the hottest part of the day.
- **Hive Placement**: The entrance of the hive should face east or southeast to allow the bees to start foraging early in the day when the sun rises.

3. Bee Care and Maintenance

- **Hive Inspections**: Regular hive inspections (every few weeks) are necessary to monitor the health of the colony, check for diseases, and ensure the bees have enough space to grow.
- **Swarming**: Bees may swarm when the hive becomes overcrowded. If you notice swarming behavior, consider adding more space to the hive or splitting the colony to prevent losing bees.
- **Pests and Diseases**: Common threats to bee colonies include varroa mites, small hive beetles, and foulbrood. Monitor your hives for signs of these pests and treat them promptly with natural remedies or hive management techniques.

4. Harvesting Honey and Beeswax

- **Honey Extraction**: Honey can be harvested once the comb is capped (sealed with beeswax). Use a honey extractor to remove honey from the combs without damaging them.
- **Beeswax Harvest**: After honey extraction, beeswax can be melted down and filtered for use in candles, cosmetics, and other products.

Advanced Beekeeping: Enhancing Honey Production and Colony Health

Beekeeping is an essential part of off-grid living, offering a sustainable source of honey, beeswax, and pollination for your crops. For those looking to optimize their beekeeping practices, this section covers advanced techniques to increase honey production, improve colony health, and manage beekeeping operations more efficiently. Whether you're looking to expand your beekeeping efforts or boost the resilience of your colonies, these strategies will help ensure long-term success.

---

1. Colony Management Strategies: Maintaining Healthy and Productive Hives

A well-managed colony is key to maximizing honey production and ensuring your bees stay healthy year-round. This section focuses on advanced colony management techniques to help you keep your hives thriving.

- **Seasonal Hive Management**: As seasons change, so do the needs of your bees. Proper seasonal management ensures that your colony remains strong and productive.
    - **Spring**: In early spring, inspect your hive to ensure the queen is laying eggs and the population is growing. This is the time to expand your hive by adding new frames and supers to accommodate the growing colony and honey stores.
    - **Summer**: During peak honey production, monitor the hive regularly for overcrowding, which can lead to swarming. Swarming occurs when the colony becomes too large, and the bees split off to form a new hive. Use **queen excluders** to prevent swarming and ensure optimal honey production.

  - **Fall**: As temperatures drop, reduce the size of the hive by removing unused supers and consolidating the colony's honey stores. This helps bees stay warm and improves their chances of survival during winter.
  - **Winter**: In colder climates, be sure to insulate your hive and reduce the entrance to minimize drafts. Monitor the hive periodically to ensure the bees have enough honey stores to last the winter.
- **Queen Management**: A strong and healthy queen is essential for a productive colony. Consider **re-queening** your hive every 1-2 years to maintain strong genetics and prevent disease. Replacing the queen periodically encourages strong egg production and reduces the risk of swarming. If your queen is aging or not laying eggs consistently, replace her with a new, healthy queen.
- **Managing Hive Space**: Providing the right amount of space inside the hive is crucial. If bees feel crowded, they may swarm, which can significantly reduce honey production. Add additional hive supers (boxes that hold frames for honey storage) when necessary to give the bees ample room to store honey and prevent overcrowding.

---

2. Increasing Honey Yield: Best Practices for Maximizing Production

Producing more honey requires optimizing every aspect of your beekeeping operation, from hive placement to nectar sources. Here are some proven strategies to increase honey yields.

- **Strategic Hive Placement**: The location of your hives plays a significant role in honey production. Place your hives in areas with abundant flowering plants, including fruit trees, clover, wildflowers, and other nectar-rich plants. Ensure the hives have access to water, and place them in locations sheltered from strong winds or extreme weather conditions.
  - **Pro Tip**: Face your hives towards the southeast to catch the morning sun, encouraging bees to become active earlier in the day.
- **Providing Supplemental Nectar Sources**: During dry periods or when local flowering plants are scarce, providing supplemental nectar or sugar water can help sustain your bees and boost honey production. Use a **bee feeder** filled with sugar water (1:1 sugar-to-water ratio) during times of nectar scarcity, but be sure to remove feeders once natural nectar sources become available again.
- **Selective Harvesting**: It's important to balance honey harvesting with the bees' need for winter stores. Only harvest honey from the **supers**, and leave enough honey in the brood box to sustain the colony through the winter. A good rule of thumb is to leave at least 60 pounds of honey for the colony, depending on your region and climate.
- **Comb Management**: Use a **comb rotation** strategy to maintain healthy honeycomb and prevent disease. Older combs can harbor diseases and pesticides, so it's important to replace old or damaged combs regularly. Rotate out old frames by removing a few each season and replacing them with new ones.

---

3. Enhancing Colony Health: Natural Methods for Disease and Pest Control

Keeping your bees healthy is essential to maintaining a productive hive. Disease and pest outbreaks can weaken or wipe out a colony, so it's critical to stay ahead of potential threats. In this section, we'll explore natural, sustainable methods for improving colony health and preventing common problems.

- **Pest Control**:
    - **Varroa Mite Management**: Varroa mites are one of the most common and destructive pests in beekeeping. Regular mite checks and treatments are necessary to keep infestations under control.
        - **Powdered Sugar Dusting**: A natural way to reduce mite populations is by dusting bees with powdered sugar. This encourages the bees to groom themselves, knocking mites off the hive. Apply powdered sugar to the frames every 10-14 days during the active season.
        - **Essential Oil Treatments**: Essential oils such as thyme, tea tree, and wintergreen can help repel mites. Add a few drops to your sugar water or directly into the hive using vapor treatments or medicated strips.
- **Disease Prevention**:
    - **American Foulbrood (AFB) Prevention**: AFB is a highly contagious bacterial disease that affects bee larvae. To prevent AFB, practice good hive hygiene by sterilizing tools and equipment, and regularly replacing old frames. If you detect AFB, isolate and treat the infected hive immediately, as the disease can spread quickly to other colonies.
    - **Chalkbrood Management**: Chalkbrood is a fungal disease that can affect bee larvae. Increasing hive ventilation can help prevent the buildup of moisture, which promotes fungal growth. You can also introduce bees with stronger genetics that show resistance to chalkbrood.
- **Boosting Immunity Through Nutrition**:
    - **Probiotic Supplements**: Just like humans, bees can benefit from probiotics to boost their immune systems. Add probiotics to your bees' diet during times of stress or after treatments to help restore gut health.
    - **Pollen Patties**: Supplementing the colony's diet with pollen patties during the early spring ensures that bees have the protein they need to raise strong larvae. Pollen patties also provide nutrition during times when natural pollen sources are scarce.

---

4. Honey Extraction Techniques: Harvesting Efficiently with Minimal Disruption

The honey extraction process can be disruptive to the bees, so it's important to use efficient techniques that minimize stress on the colony while maximizing your honey harvest.

- **Manual vs. Electric Extractors**:
    - **Manual Extractors**: These are ideal for small-scale operations and are less expensive, though they require more physical effort. Use a **tangential extractor**, which allows for honey removal from one side of the comb at a time, making it a slower but gentler process.
    - **Electric Extractors**: For larger beekeeping operations, electric extractors are more efficient, allowing you to extract honey quickly with less manual labor. A **radial extractor** spins the frames in a way that extracts honey from both sides simultaneously.
- **Uncapping Methods**: To access the honey in your frames, you need to uncap the wax covering. Here are a few uncapping methods:
    - **Uncapping Knife**: Use a heated uncapping knife to gently cut away the wax caps. Heated knives melt the wax as you cut, making the process smoother.

    - **Uncapping Fork**: For smaller operations, an uncapping fork can be used to scratch off the wax caps. While slower than a knife, it's less likely to damage the honeycomb.
- **Minimizing Bee Disruption**: During the extraction process, use **bee escapes** or **fume boards** to safely remove bees from the honey supers before harvesting. This reduces stress on the colony and prevents the bees from getting stuck in the extractor.
- **Honey Storage**: After extraction, store honey in airtight, food-grade containers. Keep it in a cool, dry place to prevent fermentation and crystallization. Honey can last indefinitely if stored properly.

---

5. Beeswax Collection and Utilization

Beeswax is a valuable byproduct of beekeeping that can be used in various ways, from making candles to waterproofing tools. Harvesting and processing beeswax properly ensures you get the most out of this versatile resource.

- **Harvesting Beeswax**: When you remove old comb or excess cappings from honey frames, save the wax. Clean the wax by melting it in a double boiler and straining it through cheesecloth to remove impurities.
- **Beeswax Uses**:
    - **Candle Making**: Beeswax is ideal for making long-lasting, natural candles. Melt the beeswax, pour it into molds, and add a wick for a simple and sustainable candle-making process.
    - **Waterproofing and Polishing**: Beeswax can be mixed with oils (like linseed or olive oil) to create a natural polish for wood and leather. It also acts as an effective waterproofing agent for tools or outdoor gear.
    - **Homemade Balms and Lotions**: Use beeswax as a base for creating homemade balms and lotions, combining it with natural oils and essential oils for a variety of skincare products.

Key Takeaways for Advanced Beekeeping:

- **Optimize Colony Management**: Manage hive space, prevent swarming, and ensure a healthy queen to maximize honey production and maintain a strong colony year-round.
- **Increase Honey Yield**: Strategic hive placement, supplemental feeding, and selective harvesting practices will help you increase honey production without compromising the health of your bees.
- **Enhance Colony Health**: Use natural methods for pest and disease control, along with nutritional supplements, to ensure your bees remain healthy and productive.
- **Efficient Honey Extraction**: Use the right tools and techniques to harvest honey efficiently, minimizing disruption to the colony while preserving honey quality.
- **Utilize Beeswax**: Collect and process beeswax for a variety of practical uses, from candle making to waterproofing and natural skincare products.

By mastering these advanced beekeeping techniques, you can enhance your honey production, keep your colonies healthy, and fully utilize the resources provided by your bees, supporting a sustainable and self-sufficient off-grid lifestyle.

11. Techniques for Food Preservation

Livestock and beekeeping contribute to a surplus of food products that can be preserved for long-term storage. This section covers traditional methods for preserving meat, dairy, and honey.

11.1 Smoking, Curing, and Dehydrating Meat

Preserving meat through smoking, curing, and dehydrating allows you to store large amounts of food without refrigeration.

- **Smoking**: Smoking meat slowly over a low heat helps preserve it and infuse it with flavor. Use hardwoods like oak or hickory for the best results.
- **Curing with Salt**: Salt curing draws out moisture and preserves meat by inhibiting bacterial growth. This method is ideal for preserving pork and beef.
- **Dehydrating**: Meat can be dried in a solar dehydrator or hung in a dry, airy place. Dehydrated meat lasts for months and is ideal for making jerky.

11.2 DIY Solar Dehydrator Construction

A solar dehydrator uses the power of the sun to dry meat, fruits, and vegetables. It's a low-cost, efficient way to preserve food for long-term storage.

- **Building a Solar Dehydrator**: A simple solar dehydrator can be constructed using wood, mesh trays, and a clear plastic cover to capture solar heat. Place the food on the trays and allow it to dry over several days.

Key Takeaways for Livestock Raising and Beekeeping:

- **Livestock Selection**: Choose hardy, adaptable livestock breeds based on your land, resources, and food needs. Chickens, pigs, and goats are excellent choices for small, off-grid homesteads.
- **Beekeeping**: Bees are low-maintenance but provide high-value products like honey, beeswax, and pollination services. Ensure your apiary is well-placed and regularly inspected.
- **Food Preservation**: Learn traditional food preservation techniques like smoking, curing, and dehydrating to extend the shelf life of your meat and dairy products.

Raising livestock and keeping bees are both sustainable ways to enhance your self-sufficiency and create a more resilient off-grid homestead. By carefully managing your animals and maintaining your beehives, you can produce a continuous supply of food, honey, and other valuable resources.

12: Advanced Gardening Techniques

A survival garden is a cornerstone of self-sufficiency for anyone living off the grid. It provides a reliable, renewable source of food and reduces your dependence on external supplies. Unlike traditional gardening, a survival garden focuses on growing nutrient-dense crops that can be preserved, stored, and relied upon in times of need. This chapter explores how to plan, plant, and maintain a sustainable survival garden, focusing on the techniques that maximize productivity while minimizing resource use.

12.1 Introduction to Off-Grid Gardening

Off-grid gardening is the foundation of self-sufficiency, enabling you to grow your own food and reduce your dependence on external resources. Unlike conventional gardening, off-grid gardening emphasizes sustainability, resource conservation, and the ability to thrive without modern conveniences such as electricity, chemical fertilizers, or automated irrigation systems. Whether you are living off the grid by choice or preparing for emergencies, cultivating a survival garden can provide you with a renewable source of nutritious food.

In this section, we'll explore the fundamentals of planning and creating a self-sufficient garden, including choosing the right location, building raised beds, and using sustainable methods that maximize productivity and efficiency.

Designing a Self-Sufficient Raised Bed Garden: Step-by-Step Instructions

Raised bed gardening is an excellent technique for off-grid living. It improves soil quality, conserves water, and maximizes productivity, even in small spaces. Raised beds also provide better drainage, minimize weeds, and allow for easy access to your plants.

1. Choosing the Location

The success of your survival garden largely depends on its location. You need to find a space that optimizes sunlight exposure, water access, and soil quality.

- **Sunlight**: Most crops require at least 6–8 hours of direct sunlight per day. Select a spot that receives full sun, especially for vegetables like tomatoes, peppers, and beans, which are sunlight-intensive.
- **Water Access**: Ensure your garden is located near a water source, whether that's a rainwater collection system, a natural spring, or a nearby pond. Easy access to water reduces the labor involved in irrigation and ensures that your crops get the hydration they need, especially during dry periods.
- **Soil Quality**: Good soil is the foundation of a successful garden. If your soil is rocky, sandy, or lacking nutrients, you can improve it by adding organic matter such as compost, manure, or leaf mulch. For raised beds, you can control the soil mixture more easily, which is one of their biggest advantages.

---

2. Building the Raised Beds

Raised beds are a simple yet effective way to boost your garden's productivity. They offer better control over soil conditions and make gardening more manageable, especially in off-grid settings where maintaining large plots of land may not be feasible.

Materials:

- **Wood**: Untreated wood is a popular and affordable choice for raised bed walls. Cedar and redwood are naturally rot-resistant and can last for several years.
- **Stone or Bricks**: For a more permanent structure, you can use stone or bricks to build your beds. These materials are more durable and require less maintenance over time.
- **Recycled Materials**: In true off-grid fashion, repurposing materials like old pallets or metal siding is an eco-friendly and cost-effective way to build raised beds.

Construction:

- **Dimensions**: A typical raised bed should be 6–12 inches deep and 4 feet wide. The width allows you to reach into the bed from either side without stepping on the soil, which prevents compaction. Length can vary based on available space, but 8–12 feet is common.
- **Filling the Beds**: Once your raised bed structure is complete, fill it with a mixture of native soil, compost, and organic material like aged manure or leaf mold. A 50/50 mix of soil and compost will provide the nutrients and drainage your plants need to thrive.

Pathways:

- **Mulching Pathways**: To create low-maintenance pathways between your beds, cover the ground with mulch such as straw, wood chips, or gravel. This suppresses weeds, retains moisture, and creates a comfortable walking surface.
- **Accessibility**: Leave at least 18–24 inches between beds for easy access with tools or wheelbarrows.

---

3. Planting the Survival Crops

A survival garden is focused on growing high-yield, nutrient-dense crops that can be stored or preserved for future use. The choice of crops should be tailored to your climate, soil, and specific needs, but many crops are universally valuable for their nutritional benefits and ease of growth.

Top Survival Crops:

- **Potatoes**: High in calories and carbohydrates, potatoes are a staple in survival gardens. They store well and can be grown in a variety of conditions.
- **Beans**: Beans are an excellent source of protein and fiber. They can be dried for long-term storage and are easy to grow in most climates.
- **Cabbage and Kale**: Leafy greens like cabbage and kale provide essential vitamins and minerals. Kale, in particular, is hardy and can grow in cold climates, making it a reliable food source year-round.

- **Squash**: Squash, both summer and winter varieties, produce large yields and can be stored for months without refrigeration.
- **Carrots and Beets**: Root vegetables like carrots and beets are nutrient-dense and store well through the winter in root cellars or cold storage.

Companion Planting:

Companion planting is a sustainable gardening technique that involves growing certain plants together for mutual benefit. This method helps improve yields, reduce pests, and enrich the soil naturally.

- **Examples of Companion Plants**:
    - **Beans and Corn**: Beans fix nitrogen in the soil, which benefits corn. In return, the corn stalks provide support for climbing bean plants.
    - **Tomatoes and Basil**: Basil can help repel pests that commonly attack tomatoes, while improving the flavor of the tomatoes.
    - **Carrots and Onions**: Onions help deter pests that target carrots, making them an excellent companion pair.

Succession Planting:

To maximize the productivity of your raised beds, practice succession planting. This involves planting new crops immediately after harvesting the previous ones, ensuring continuous production throughout the growing season.

- **Cool-Season Crops**: Start with cool-season crops like lettuce, peas, and spinach in early spring.
- **Warm-Season Crops**: Once these are harvested, replace them with warm-season crops like beans, tomatoes, and peppers.
- **Late-Season Crops**: In late summer, plant fall crops such as broccoli, kale, and radishes to extend your harvest into the cooler months.

Watering and Irrigation in Off-Grid Gardens

In off-grid environments, you must conserve water while ensuring that your plants get enough moisture to thrive. Effective irrigation is crucial, and there are several methods to achieve this in a sustainable way.

1. Rainwater Harvesting:

Collecting rainwater is one of the most efficient ways to irrigate your garden without relying on external water sources. Install gutters and downspouts on your home or garden structures to direct rainwater into barrels or tanks. Use this stored water to irrigate your crops during dry periods.

2. Drip Irrigation:

Drip irrigation systems are ideal for off-grid gardens because they deliver water directly to the plant's roots, minimizing evaporation and runoff. You can set up a simple gravity-fed drip system using rainwater barrels and plastic tubing.

3. Mulching:

Mulching helps retain moisture in the soil by reducing evaporation. Organic mulches like straw, leaves, or wood chips also improve soil health as they break down, providing additional nutrients for your crops.

Key Takeaways for Off-Grid Gardening:

- **Raised beds** offer numerous benefits, including improved soil quality, better water management, and easier access to your crops. They are a practical solution for off-grid gardens where space and resources may be limited.
- **Companion planting** and **succession planting** are sustainable gardening practices that can increase yields, reduce pests, and ensure a continuous food supply throughout the growing season.
- **Water conservation** is critical in off-grid gardening. Use rainwater harvesting, drip irrigation, and mulching to make the most of your water resources.
- **Survival crops** such as potatoes, beans, leafy greens, and root vegetables should be prioritized for their high yields, nutritional value, and ability to store or preserve for long periods.

By carefully planning and implementing sustainable gardening practices, you can create a productive off-grid garden that ensures a reliable supply of food throughout the year.

12.2 Mini-Greenhouses and Vertical Gardens

In off-grid gardening, maximizing space and extending the growing season are key to ensuring a consistent and sustainable food supply. Mini-greenhouses and vertical gardens are practical solutions for gardeners with limited space or harsh growing conditions. These systems allow you to grow more food in less space and protect your crops from extreme weather, making them indispensable tools for off-grid living.

In this section, we'll explore how to build and maintain both mini-greenhouses and vertical gardens, providing step-by-step guides for each method and discussing which crops thrive in these environments.

Building a Vertical Garden Using Pallets: Step-by-Step Guide

Vertical gardening is an efficient way to grow crops in limited spaces by stacking plants vertically instead of spreading them out horizontally. This technique is especially useful for small homesteads or urban off-grid gardens, where space is at a premium. Vertical gardens are also easier to maintain, allowing you to harvest and tend to your plants without excessive bending or weeding.

1. Choosing the Right Pallets and Materials

Using recycled wooden pallets is an affordable and sustainable way to create a vertical garden. However, it's important to choose safe materials that won't harm your plants.

- **Pallet Selection**: Use heat-treated pallets that are marked with "HT" (heat-treated) to avoid any chemical contamination. Avoid pallets that are treated with chemicals or show signs of wear and mold.
- **Additional Materials**: You'll need landscape fabric or burlap to line the inside of the pallet, soil, and organic compost to fill the planting spaces. Screws, nails, or a staple gun will help secure the fabric and ensure the soil stays in place.

2. Preparing the Pallets for Gardening

Once you've chosen the right pallet, the next step is to prepare it for planting. Proper preparation ensures the stability of your vertical garden and prevents soil from falling out.

- **Clean and Sand the Pallet**: Start by cleaning the pallet thoroughly to remove any dirt or debris. Sand down any rough edges to prevent splinters.
- **Attach the Landscape Fabric**: Line the back and sides of the pallet with landscape fabric or burlap. Use a staple gun to secure it, ensuring there are no gaps where soil could escape.
- **Fill with Soil**: Once the fabric is secure, fill the pallet with a nutrient-rich mix of soil and compost. Be sure to pack the soil tightly to avoid it settling too much once the plants are added.

3. Planting Your Vertical Garden

Now that your pallet is prepared, it's time to plant. Choose crops that do well in vertical settings and have shallow root systems.

- **Best Crops for Vertical Gardens**:
  - **Leafy Greens**: Lettuce, spinach, and arugula thrive in vertical gardens. These crops are light, have shallow roots, and can be harvested continuously throughout the season.

  - **Herbs**: Basil, parsley, cilantro, and thyme grow well in compact spaces and add flavor to your meals.
  - **Strawberries**: Strawberries are ideal for vertical gardens, as they trail downward and benefit from the elevated structure, which keeps the fruit off the ground and reduces the risk of rot.
  - **Peas and Beans**: Climbing plants like peas and beans are perfect for vertical gardens, as they can use the pallet's structure for support as they grow.
- **Planting Method**: Insert the plants or seeds into the soil pockets between the wooden slats of the pallet. Water the plants thoroughly after planting, ensuring the soil is evenly moist.

4. Maintaining Your Vertical Garden

Vertical gardens require slightly different maintenance than traditional gardens. Proper watering and structural support are key to keeping your plants healthy.

- **Watering**: Vertical gardens tend to dry out faster than ground-level beds, so you'll need to water them more frequently. Consider installing a drip irrigation system or manually water the plants in the early morning or late afternoon to prevent water evaporation.
- **Fertilizing**: Because vertical gardens contain less soil, they may require more frequent fertilization. Use organic compost or a liquid fertilizer every few weeks to replenish nutrients.
- **Pruning**: Regularly prune and harvest your plants to encourage continued growth and prevent overcrowding in the pallet.

Simple Greenhouse and Vertical Gardening Ideas

In addition to vertical gardens, mini-greenhouses offer an effective way to extend the growing season and protect your crops from adverse weather conditions. Greenhouses trap heat from the sun, allowing you to grow plants earlier in the spring and later into the fall.

1. Building a Simple Mini-Greenhouse

A mini-greenhouse can be as simple or elaborate as you want, depending on the materials available and your specific needs. Even a basic structure can significantly extend your growing season and improve yields.

Materials:

- **PVC Pipes or Wood**: PVC pipes are lightweight, affordable, and easy to work with. You can also use wood if you prefer a more permanent structure.
- **Clear Plastic Sheeting**: This serves as the greenhouse's cover, allowing sunlight to enter while trapping heat inside.
- **Fasteners**: Use zip ties, clamps, or screws to attach the plastic sheeting to the frame.

Construction:

- **Frame**: Create a frame using PVC pipes or wooden boards, ensuring that the structure is sturdy and large enough to cover your garden bed or a section of your garden.
- **Plastic Cover**: Drape the clear plastic sheeting over the frame, securing it tightly at the corners with fasteners. Ensure that the cover is taut to prevent it from blowing away in strong winds.
- **Ventilation**: Leave one side of the greenhouse open or create a flap that can be lifted to allow for airflow. Proper ventilation is essential to prevent overheating and ensure healthy plant growth.

Benefits of a Mini-Greenhouse:

- **Extended Growing Season**: Mini-greenhouses allow you to start seeds earlier in the season and grow crops later into the fall or even winter in some climates.
- **Protection from Weather**: The greenhouse protects plants from frost, heavy rain, and high winds, providing a more stable growing environment.

2. Cold Frames

Cold frames are another simple and effective way to extend the growing season. These low-lying structures act as mini-greenhouses, trapping solar heat and insulating plants from cold temperatures.

Building a Cold Frame:

- **Materials**: You can build a cold frame using wood or bricks for the frame and an old window or clear plastic for the top.
- **Construction**: Build a box-like structure with slanted sides. The top should be hinged so you can open it during the day for ventilation. The slanted design allows the cold frame to capture maximum sunlight.
- **Positioning**: Place your cold frame in a sunny spot where it will receive plenty of light during the day. The slanted top should face south to capture the most sunlight.

Choosing Crops for Greenhouses and Vertical Gardens

Not all crops are suitable for greenhouses or vertical gardens, so it's important to select plants that will thrive in these environments. Greenhouses provide warmth and protection, while vertical gardens maximize growing space for smaller crops.

Best Crops for Greenhouses:

- **Tomatoes**: Greenhouses provide the warmth and humidity that tomatoes need to thrive. They also protect tomatoes from pests like hornworms and diseases such as blight.
- **Peppers**: Peppers love the heat, and greenhouses provide the perfect conditions for growing them year-round.
- **Cucumbers**: Cucumbers grow well in the controlled environment of a greenhouse, where they are less likely to suffer from mildew and rot.
- **Herbs**: Herbs such as basil, mint, and cilantro grow quickly and benefit from the warmth of a greenhouse.

Best Crops for Vertical Gardens:

- **Leafy Greens**: Lettuce, spinach, and arugula are lightweight and thrive in vertical gardens where space is limited.
- **Strawberries**: These trailing plants grow beautifully in vertical gardens, and the elevated structure helps prevent fruit rot.
- **Climbing Plants**: Beans, peas, and other climbing plants can take full advantage of vertical spaces, using the structure of the garden to grow upward.

Key Takeaways for Mini-Greenhouses and Vertical Gardens:

- **Vertical gardens** maximize space and allow you to grow crops even in small or urban areas. Pallets provide an affordable and efficient structure for creating a vertical garden.
- **Mini-greenhouses** and **cold frames** extend the growing season by protecting crops from frost, cold weather, and harsh environmental conditions.
- **Sustainable crops** such as leafy greens, strawberries, and climbing plants thrive in vertical gardens, while heat-loving crops like tomatoes and peppers benefit from the warmth of greenhouses.
- **Watering and maintenance** are key in both mini-greenhouses and vertical gardens, as they can dry out quickly. Regular watering, fertilization, and pruning will ensure healthy plant growth.

By integrating mini-greenhouses and vertical gardens into your off-grid gardening plan, you can maximize productivity, extend your growing season, and make the most of limited space. These methods are particularly valuable for those living in areas with harsh climates or limited resources.

12.3 Pest Control and Sustainable Cultivation

One of the main challenges in off-grid gardening is managing pests and diseases without access to synthetic chemicals or commercial pesticides. Sustainable pest control methods are essential for maintaining a healthy garden that produces abundant food without damaging the environment or relying on non-renewable resources. In this section, we'll explore natural, organic techniques for controlling pests and preventing crop diseases while promoting sustainable cultivation practices.

Organic Pest Control Techniques and Crop Disease Management

To protect your crops and ensure a successful harvest, you'll need to implement natural, sustainable pest control methods. These methods help maintain the ecological balance of your garden while keeping harmful insects and diseases at bay. The key is to use preventive techniques and maintain healthy soil, which strengthens plants and makes them less vulnerable to pests and diseases.

1. Companion Planting for Pest Control

Companion planting is a time-tested method of pest control that involves growing certain plants together to improve growth and repel pests. By selecting the right combinations of plants, you can naturally reduce pest populations and promote healthier crops.

- **Marigolds**: Plant marigolds around your vegetable beds. Their strong scent repels nematodes, aphids, and mosquitoes, while attracting beneficial insects like ladybugs and hoverflies, which feed on pests.
- **Basil and Tomatoes**: Basil planted near tomatoes can help repel whiteflies, aphids, and mosquitoes. In addition to keeping pests away, basil improves the flavor of tomatoes.
- **Garlic and Onions**: These alliums naturally repel a wide range of insects, including aphids, carrot flies, and cabbage worms. Plant them near carrots, tomatoes, or lettuce to protect these crops.
- **Nasturtiums**: These flowers attract aphids away from more valuable crops, acting as a "sacrificial plant" to protect your vegetables.

Companion planting enhances biodiversity and helps maintain the natural balance of your garden. It also encourages beneficial insects that act as predators to harmful pests.

2. Homemade Natural Pesticides

In off-grid environments, you can create effective, organic pesticides using common household ingredients. These natural solutions are safe for your plants and the environment, and they help control pests without the use of chemicals.

- **Soap and Water Spray**: A mixture of water and mild soap (such as castile soap) can be sprayed directly onto plants to eliminate soft-bodied insects like aphids, spider mites, and whiteflies. Simply mix 2 tablespoons of soap in 1 quart of water and spray onto the affected plants every few days.
- **Neem Oil**: Neem oil is a powerful, all-natural pesticide that targets a wide range of insects, including caterpillars, beetles, and mites. It also acts as a fungicide, preventing mildew and fungal infections on your crops. Dilute the oil according to the manufacturer's instructions and apply to plants using a spray bottle.
- **Garlic and Chili Spray**: Garlic and chili pepper spray is an effective, organic insect repellent. Blend a few garlic cloves and chili peppers with water, strain the mixture, and

spray it onto your plants. This natural spray will repel a variety of pests, including aphids and beetles.
- **Diatomaceous Earth**: This powder, made from the fossilized remains of diatoms, is a natural insect deterrent. When sprinkled around the base of plants, it acts as a barrier that repels slugs, snails, and crawling insects like ants and earwigs.

3. Attracting Beneficial Insects

Another important aspect of natural pest control is encouraging beneficial insects that prey on harmful pests. By creating an environment that attracts these helpful creatures, you can naturally reduce pest populations.

- **Ladybugs**: These insects feed on aphids, mites, and other small pests. Planting herbs like dill, fennel, and cilantro will attract ladybugs to your garden.
- **Lacewings**: Green lacewings are another valuable predator of aphids and caterpillars. Planting sunflowers, cosmos, and yarrow can attract these beneficial insects.
- **Hoverflies**: Hoverflies pollinate flowers and feed on aphids, thrips, and other soft-bodied insects. They are drawn to gardens with flowering plants such as marigolds, daisies, and alyssum.
- **Bees**: While not predators, bees are essential for pollinating your crops. Plant a variety of bee-friendly flowers such as lavender, sunflowers, and wildflowers to encourage pollination.

Disease Management in Off-Grid Gardening

Managing plant diseases is equally important in off-grid gardening, as certain diseases can devastate crops if left unchecked. Disease management focuses on prevention and early detection, ensuring that your plants stay healthy throughout the growing season.

1. Preventing Plant Diseases

Prevention is the best defense against plant diseases, especially in environments where you cannot rely on commercial fungicides or treatments. Here are some key strategies to prevent the spread of disease in your garden:

- **Crop Rotation**: Rotate your crops each year to prevent soil-borne diseases from building up. For example, avoid planting members of the same plant family (such as tomatoes and peppers) in the same spot for consecutive years.
- **Proper Spacing**: Ensure that plants are spaced adequately to allow for good air circulation. Overcrowded plants are more susceptible to fungal diseases like powdery mildew and blight.
- **Watering at the Base**: Water your plants at the base rather than from overhead to prevent wetting the leaves. Wet leaves are more likely to develop fungal infections.
- **Sanitize Tools**: Clean and disinfect your gardening tools between uses to avoid spreading disease from one plant to another.

2. Managing Common Plant Diseases

Even with the best preventive measures, diseases can sometimes affect your crops. Being able to identify and manage these diseases quickly is critical for saving your plants.

- **Powdery Mildew**: Powdery mildew is a common fungal infection that causes a white powdery coating on plant leaves. To prevent powdery mildew, avoid overwatering and ensure your plants are spaced out for proper airflow. If powdery mildew appears, use a spray made from 1 part milk to 9 parts water to treat the affected areas.
- **Blight (Early or Late)**: Blight affects tomatoes, potatoes, and other members of the nightshade family. Early detection is key, so regularly inspect your plants for brown spots on the leaves or stems. Remove any infected leaves and improve air circulation around the plants. Copper-based fungicides, applied early, can help control blight.
- **Root Rot**: Caused by overly wet soil, root rot can kill plants quickly. To prevent this, ensure that your garden beds or containers have good drainage. If you notice yellowing leaves and stunted growth, dig up the plant to check for rotting roots. Remove affected plants and improve soil drainage to prevent the disease from spreading.
- **Rust**: Rust is a fungal disease that causes reddish-brown spots on leaves, particularly in beans and garlic. Prevent rust by watering plants early in the morning and removing infected leaves as soon as they are spotted. Neem oil can also be used to control the spread of rust.

Sustainable Soil Management for Long-Term Success

Healthy soil is the foundation of a thriving garden, and sustainable soil management is key to ensuring long-term productivity. By building and maintaining healthy soil, you reduce the risk of diseases, improve water retention, and provide your plants with the nutrients they need to grow strong.

1. Composting and Mulching

Composting and mulching are two simple, effective ways to improve soil fertility and structure while conserving water.

- **Composting**: Compost is rich in organic matter and nutrients, making it one of the best natural fertilizers for your garden. Start a compost pile with kitchen scraps, grass clippings, leaves, and manure. Turn the compost regularly to speed up decomposition, and add the finished compost to your garden beds to improve soil health.
- **Mulching**: Organic mulches, such as straw, wood chips, or leaves, help retain soil moisture, suppress weeds, and add organic matter to the soil as they decompose. Apply a thick layer of mulch around your plants to protect the soil from drying out and prevent erosion.

2. Cover Crops and Green Manure

Planting cover crops, also known as green manure, is an excellent way to improve soil fertility and structure. Cover crops protect the soil from erosion and, when tilled into the soil, add valuable nutrients back into the ground.

- **Legumes**: Crops like clover, alfalfa, and peas fix nitrogen in the soil, enriching it for future plantings.

- **Grasses and Grains**: Oats, rye, and wheat can be grown as cover crops to prevent soil erosion and add organic matter to the soil when tilled under.

Key Takeaways for Pest Control and Sustainable Cultivation:

- **Companion planting** and **attracting beneficial insects** are natural, sustainable methods of pest control that reduce the need for chemical pesticides while promoting biodiversity in your garden.
- **Homemade natural pesticides** like soap sprays, neem oil, and garlic chili spray are effective, eco-friendly solutions for controlling common pests.
- **Crop rotation**, **proper watering techniques**, and **good air circulation** are essential for preventing and managing plant diseases without synthetic fungicides.
- **Sustainable soil management** through composting, mulching, and planting cover crops ensures that your garden remains productive and resilient for the long term.

By implementing these organic pest control and disease management strategies, along with sustainable soil practices, you can maintain a healthy, thriving garden that produces abundant food while preserving the health of the ecosystem.

Building a Permaculture Garden

A **permaculture garden** is designed to mimic natural ecosystems, creating a sustainable, self-sufficient food production system that requires minimal external inputs. By incorporating principles of biodiversity, soil health, and water conservation, permaculture gardens can produce an abundance of food while regenerating the environment. In this section, we'll explore how to design and build a permaculture garden tailored to off-grid living, ensuring long-term sustainability and resilience.

---

1. Principles of Permaculture: Designing for Sustainability

Permaculture is based on a set of principles that guide the design and management of a sustainable ecosystem. Understanding and applying these principles ensures that your garden works in harmony with nature, rather than against it.

- **Observe and Interact**: Take time to observe your land before starting your garden. Understand the natural cycles, such as the sun's path, wind patterns, and water flow. This information will help you design your garden in a way that maximizes natural resources.
- **Catch and Store Energy**: In a permaculture system, capturing and storing energy is key. This could mean capturing rainwater in tanks or ponds, using solar power to run garden equipment, or storing nutrients in compost. By storing these resources, you can make them available during dry spells or when other inputs are limited.
- **Integrate Rather Than Segregate**: A core principle of permaculture is to integrate different elements of your garden to work together symbiotically. For example, plant herbs near fruit trees to repel pests or grow nitrogen-fixing plants (such as legumes) near heavy feeders to enrich the soil naturally.
- **Use Renewable Resources**: Focus on creating a self-sustaining garden that relies on renewable resources like solar energy, composting, and natural pest control. Avoid chemical fertilizers and pesticides, opting instead for organic alternatives.
- **Use and Value Diversity**: A diverse garden mimics natural ecosystems and is more resilient to pests, diseases, and changing weather conditions. Incorporate a wide variety of plants, animals, and insects into your system to create a balanced, thriving ecosystem.

---

2. Edible Forest Gardens: Mimicking Natural Ecosystems

An **edible forest garden** is a permaculture concept that mimics a natural forest ecosystem, with different layers of vegetation working together to create a self-sustaining food source. The garden consists of various layers, from tall fruit and nut trees to ground cover plants, all contributing to soil health and biodiversity.

- **Layers of an Edible Forest Garden**:
  - **Canopy Layer**: The tallest layer, made up of fruit and nut trees like apple, pear, and chestnut. These trees provide shade and fruit, while their deep roots help to stabilize the soil.
  - **Understory Layer**: Smaller trees and shrubs, such as dwarf fruit trees, hazelnuts, and berry bushes (raspberries, currants). These plants thrive in partial shade and provide additional yields.
  - **Herbaceous Layer**: This layer includes herbs and perennial vegetables like mint, oregano, rhubarb, and asparagus. They fill in the gaps between trees and shrubs, helping to prevent erosion and suppress weeds.
  - **Ground Cover Layer**: Plants like clover, strawberries, and creeping thyme make up this layer. Ground covers help protect the soil, fix nitrogen, and reduce the need for weeding.
  - **Root Layer**: Root crops such as garlic, onions, carrots, and potatoes are planted here. These plants draw nutrients from deep in the soil, ensuring your garden is productive from top to bottom.
  - **Vines and Climbers**: Use climbing plants like grapes, beans, and cucumbers to make use of vertical space. These can grow on trees or trellises, maximizing production without taking up ground space.
- **Designing an Edible Forest Garden**:

- **Start with the Canopy**: Begin by planting your tallest trees, ensuring they are spaced far enough apart to allow sunlight to reach lower layers. Consider the mature size of the trees to avoid overcrowding.
- **Add Layers Gradually**: As the trees grow, begin planting the understory, herbaceous, and ground cover layers. This approach allows you to slowly build a diverse and resilient ecosystem.
- **Mulching and Soil Building**: Cover the soil around your plants with natural mulch like straw, leaves, or wood chips to retain moisture, suppress weeds, and build fertile soil over time.

---

3. Water Conservation Techniques: Capturing and Storing Water

Water is a critical resource for any garden, and in a permaculture system, the goal is to capture and store as much water as possible, minimizing waste. Here are several techniques to manage water sustainably in your garden.

- **Swales**: Swales are shallow ditches dug on contour lines across the landscape, designed to capture rainwater and allow it to slowly seep into the soil. Swales prevent water runoff and erosion while ensuring that your plants have a steady supply of moisture. To build a swale:
    - Dig a trench on a gentle slope, following the natural contour of the land.
    - Fill the trench with organic material like wood chips or straw to slow down water flow.
    - Plant trees or shrubs along the swale to take advantage of the water it collects.
- **Hugelkultur Beds**: Hugelkultur is a method of building raised garden beds on top of rotting wood. The wood holds water like a sponge, reducing the need for irrigation and creating rich, fertile soil as it decomposes.
    - Start by digging a shallow trench and filling it with logs, branches, and other organic matter.
    - Cover the wood with soil, and plant your crops on top. Over time, the wood will break down, releasing nutrients and retaining water.
- **Rainwater Harvesting**: Collecting rainwater from rooftops and storing it in barrels or tanks is an effective way to conserve water in a permaculture garden. Use rainwater for irrigation, and connect overflow systems to swales or ponds to maximize water retention on your land.
- **Drip Irrigation**: Drip irrigation delivers water directly to the roots of your plants, reducing evaporation and water waste. It's an ideal system for dry climates or areas with limited water resources. Use a gravity-fed drip system powered by your rainwater collection to ensure a sustainable water supply.

---

4. Soil Regeneration: Building Fertility Naturally

Healthy soil is the foundation of a productive permaculture garden. Instead of relying on chemical fertilizers, permaculture emphasizes regenerative practices that build soil fertility over time.

- **Composting**: Composting organic material (kitchen scraps, garden waste, manure) is one of the best ways to enrich your soil. Incorporate finished compost into your planting beds to add nutrients, improve soil structure, and increase water retention.

- **Hot Composting**: For a faster composting process, create a hot compost pile by layering green (nitrogen-rich) and brown (carbon-rich) materials and turning the pile regularly. This method can produce finished compost in as little as 3-6 months.
- **Vermicomposting**: Worms can also be used to break down organic material. Vermicompost (worm castings) is rich in nutrients and beneficial microbes, making it an excellent soil amendment.

- **Cover Cropping**: Planting cover crops like clover, alfalfa, or rye during the off-season helps protect your soil from erosion, adds organic matter, and fixes nitrogen in the soil. These crops can be turned under to enrich the soil before the next planting season.
- **Mulching**: Mulch not only conserves water but also builds soil fertility as it breaks down. Use organic mulches like straw, leaves, or wood chips around your plants to protect the soil, reduce evaporation, and suppress weeds.
- **Natural Fertilizers**: Instead of chemical fertilizers, use natural amendments like compost tea, worm castings, or animal manures to feed your plants. These organic fertilizers release nutrients slowly, promoting steady plant growth without harming soil microbes.

---

5. Companion Planting: Maximizing Productivity and Reducing Pests

Companion planting is the practice of growing certain plants together to enhance growth, deter pests, and maximize space. In a permaculture garden, companion planting is essential for creating a balanced and productive ecosystem.

- **Nitrogen-Fixing Plants**: Legumes (peas, beans, clover) naturally fix nitrogen from the atmosphere into the soil, enriching it for nearby plants. Plant them alongside heavy feeders like tomatoes or corn.
- **Pest-Repelling Plants**: Some plants, like marigolds, garlic, and basil, naturally repel pests. Plant them near crops vulnerable to pests to reduce the need for chemical pesticides.
- **Pollinator Attractors**: Attract bees and other pollinators by planting flowers like lavender, sunflowers, and borage. This helps ensure good pollination for fruit and vegetable plants, increasing yields.
- **Three Sisters Method**: A classic example of companion planting is the **Three Sisters Method**, where corn, beans, and squash are grown together. The corn provides support for the beans, the beans fix nitrogen in the soil, and the squash covers the ground, suppressing weeds and conserving moisture.

Key Takeaways for Building a Permaculture Garden:

- **Design for Sustainability**: Apply permaculture principles to create a garden that works with nature, building soil fertility, conserving water, and promoting biodiversity.
- **Create an Edible Forest Garden**: Layer your plants to mimic a natural forest, with tall trees, shrubs, herbs, and ground covers working together to maximize productivity and resilience.
- **Capture and Store Water**: Use swales, rainwater harvesting, and drip irrigation to conserve water and ensure your garden remains productive even during dry periods.
- **Regenerate Soil Naturally**: Use compost, cover crops, and organic mulches to build soil fertility over time, avoiding the need for chemical fertilizers.

- **Companion Planting**: Maximize your garden's productivity and health by growing plants that support each other, repel pests, and attract pollinators.

By implementing these permaculture principles and techniques, you can build a productive, self-sustaining garden that requires minimal external inputs and supports your off-grid self-sufficiency goals.

Organic Pest and Disease Control

Managing pests and diseases in a sustainable and eco-friendly way is essential for maintaining a healthy and productive permaculture garden. By using organic methods, you can protect your plants without relying on harmful chemicals that may damage the environment or the delicate balance of your garden ecosystem. In this section, we'll explore a variety of natural and organic pest and disease control techniques, including biological controls, companion planting, and homemade organic solutions.

---

1. Biological Pest Control: Using Nature to Your Advantage

One of the most effective and environmentally friendly ways to control pests in your garden is by introducing natural predators that feed on common pests. Biological pest control involves harnessing the power of beneficial insects, birds, and other organisms to maintain balance in your garden.

- **Beneficial Insects**: Attracting or introducing beneficial insects into your garden can help keep pest populations in check.
  - **Ladybugs**: These helpful beetles feed on aphids, mealybugs, and other soft-bodied insects. Encourage ladybugs by planting nectar-rich flowers such as dill, fennel, and marigold.
  - **Lacewings**: Lacewing larvae are voracious predators of aphids, thrips, and mites. They can be introduced to your garden as eggs, which will hatch and begin feeding on pests.
  - **Praying Mantises**: These large, predatory insects eat a variety of pests, including beetles, caterpillars, and grasshoppers. You can purchase praying mantis egg cases to place in your garden.
  - **Nematodes**: Beneficial nematodes are microscopic worms that feed on soil-dwelling pests like grubs, rootworms, and termites. They are safe for plants and animals but highly effective against pests that damage plant roots.
- **Attracting Birds**: Birds can help control pests like caterpillars, beetles, and grasshoppers. Install birdhouses, bird feeders, and birdbaths to encourage insectivorous birds like wrens, chickadees, and swallows to visit your garden.
- **Frogs and Toads**: These amphibians are great for controlling slugs, snails, and other garden pests. Create a moist, shady area with a small pond or water feature to attract frogs and toads to your garden.

---

2. Companion Planting for Pest Control

Companion planting is a natural way to deter pests by using the repellent properties of certain plants. By planting pest-repelling plants alongside vulnerable crops, you can minimize damage without using chemical pesticides.

- **Pest-Repelling Plants**:
    - **Marigolds**: Plant marigolds near tomatoes, peppers, and cucumbers to repel aphids, nematodes, and whiteflies. The strong scent of marigolds also deters rabbits and deer.
    - **Basil**: Basil repels mosquitoes, flies, and thrips. Plant basil near tomatoes to boost growth and reduce pest pressure.
    - **Garlic and Onions**: These alliums have natural antifungal and antibacterial properties. Plant garlic and onions near roses, carrots, and lettuce to repel aphids, beetles, and caterpillars.
    - **Lavender**: The strong scent of lavender deters moths, fleas, and flies. Plant lavender around garden borders to protect crops from pests.
- **Trap Cropping**: Use trap crops to lure pests away from your main crops. Planting a sacrificial crop that pests are particularly attracted to can help keep them away from your more valuable plants.
    - **Example**: Planting nasturtiums near brassicas (like cabbage and broccoli) can attract aphids and caterpillars to the nasturtiums, sparing your main crops from damage.
- **Three Sisters Planting**: This ancient method involves planting corn, beans, and squash together. Corn provides support for beans, beans fix nitrogen in the soil, and squash covers the ground, suppressing weeds and deterring pests with its prickly leaves.

---

3. Homemade Organic Pesticides

There are many simple, homemade solutions that can effectively control pests without harming the environment. These organic pesticides are easy to make using common household ingredients.

- **Neem Oil Spray**: Neem oil is a natural pesticide derived from the neem tree. It disrupts the life cycle of insects, preventing them from feeding, molting, or laying eggs. To make a neem oil spray:
    - Mix 2 teaspoons of neem oil with 1 teaspoon of liquid soap in 1 quart of water.
    - Spray on affected plants, focusing on the undersides of leaves where pests tend to hide.
    - Neem oil is effective against aphids, spider mites, whiteflies, and caterpillars.
- **Garlic and Chili Pepper Spray**: This potent spray repels a wide variety of pests, including aphids, beetles, and caterpillars.
    - Blend 10 cloves of garlic, 1 tablespoon of chili powder, and 1 quart of water. Strain the mixture and add 1 teaspoon of liquid soap.
    - Spray the solution on plants to deter pests.
- **Diatomaceous Earth**: Diatomaceous earth is a fine powder made from fossilized algae. It works by dehydrating insects and is effective against slugs, beetles, and ants.
    - Sprinkle diatomaceous earth around the base of plants or directly on pests. Be sure to reapply after rain, as it loses effectiveness when wet.
- **Soap and Water Spray**: A simple soap and water mixture can suffocate soft-bodied insects like aphids and whiteflies.

    - Mix 1 tablespoon of liquid dish soap with 1 quart of water and spray directly on pests. The soap breaks down the insect's outer layer, causing them to dehydrate and die.
- **Baking Soda Fungicide**: Baking soda is effective against fungal diseases like powdery mildew and black spot.
    - Mix 1 tablespoon of baking soda, 1 teaspoon of liquid soap, and 1 quart of water. Spray on affected plants to prevent and treat fungal infections.

---

4. Organic Disease Control Techniques

Preventing and controlling plant diseases is just as important as managing pests. Organic disease control focuses on improving plant health and using natural methods to prevent the spread of pathogens.

- **Crop Rotation**: Rotating your crops each season helps prevent soil-borne diseases from building up in the soil. By changing the location of different plant families, you disrupt the life cycle of pathogens.
    - **Example**: Avoid planting tomatoes in the same spot where potatoes or peppers were grown the previous year to prevent soil-borne diseases like blight.
- **Improving Air Circulation**: Many fungal diseases thrive in humid, crowded conditions. Prune plants regularly to improve air circulation, and space plants adequately to prevent overcrowding.
- **Compost Tea**: Compost tea is a nutrient-rich liquid made from soaking compost in water. It contains beneficial microbes that can help fight off plant diseases. Apply compost tea as a foliar spray or soil drench to boost plant health and immunity.
- **Mulching for Disease Prevention**: Organic mulch, such as straw or wood chips, helps prevent soil-borne pathogens from splashing onto plants during watering or rain. Mulch also retains soil moisture and suppresses weeds, further promoting plant health.
- **Removing Infected Plants**: If a plant becomes infected with a disease, remove it immediately to prevent the spread to other plants. Dispose of the infected plant material away from your garden to avoid contaminating your compost.

---

5. Integrated Pest Management (IPM): A Holistic Approach

Integrated Pest Management (IPM) combines multiple strategies to manage pests and diseases in a sustainable way. The goal of IPM is not to eliminate pests entirely but to keep their populations at manageable levels while minimizing harm to the environment.

- **Monitoring and Identification**: Regularly inspect your plants for signs of pests or disease. Early detection is key to preventing outbreaks. Use sticky traps or pheromone traps to monitor insect populations.
- **Cultural Controls**: Implement cultural practices like crop rotation, intercropping, and proper irrigation to prevent pests and diseases from becoming established in your garden.
- **Mechanical Controls**: Use physical methods like handpicking pests, setting up barriers (such as row covers or netting), and installing traps to reduce pest populations without using chemicals.

- **Biological and Chemical Controls**: When necessary, use biological controls (beneficial insects) and organic pesticides to target specific pests. Only apply treatments when pest populations reach a damaging threshold to minimize the impact on beneficial organisms.

Key Takeaways for Organic Pest and Disease Control:

- **Biological Control**: Encourage beneficial insects, birds, and amphibians to naturally control pest populations.
- **Companion Planting**: Use plants that repel pests or attract beneficial insects to protect your crops.
- **Homemade Pesticides**: Create your own organic pest control sprays using neem oil, garlic, and chili peppers to keep pests at bay.
- **Disease Prevention**: Use crop rotation, proper spacing, and natural remedies like compost tea to prevent and control plant diseases.
- **Integrated Pest Management (IPM)**: Adopt a holistic approach to pest and disease control by combining multiple organic strategies that minimize environmental impact.

By integrating these organic pest and disease control methods into your permaculture garden, you can maintain a healthy and productive ecosystem without relying on synthetic chemicals. This sustainable approach will protect your crops while supporting the long-term health of your garden.

14. Off-Grid Hygiene and Health

Maintaining proper hygiene and health is crucial for survival, especially in an off-grid setting where access to modern healthcare facilities and sanitation systems may be limited. Off-grid living requires practical solutions for personal hygiene, waste management, first aid, and even natural remedies to prevent illness and maintain well-being. In this chapter, we will explore sustainable, DIY solutions for keeping yourself and your living environment clean and healthy, while also addressing how to manage waste and prepare for medical emergencies.

14.1 Personal Hygiene in Off-Grid Settings

In off-grid living, personal hygiene becomes a critical factor in preventing disease and maintaining overall health. Without access to conventional plumbing, you need to adopt alternative methods for keeping yourself clean and sanitary.

DIY Toothpaste and Natural Soap Recipes

In an off-grid environment, making your own hygiene products from simple, natural ingredients ensures that you stay clean without relying on store-bought supplies.

DIY Toothpaste Recipe

This homemade toothpaste is easy to make and uses only natural, non-toxic ingredients.

- **Ingredients**:
  - 2 tablespoons of baking soda (a natural cleanser and whitener)
  - 2 tablespoons of coconut oil (which has antibacterial properties)
  - 10 drops of peppermint essential oil (for a fresh flavor and breath)
- **Instructions**:

0. Mix the baking soda and coconut oil in a small jar until they form a paste.
1. Add the peppermint essential oil and stir well.
2. Store the paste in a sealed container and use as regular toothpaste.

DIY Natural Soap Recipe

Making your own soap ensures you have a constant supply of a necessary hygiene product, even off the grid.

- **Ingredients**:
  - 16 oz of olive oil (or any other vegetable oil)
  - 4 oz of coconut oil (for added lather)
  - 2 oz of lye (sodium hydroxide, which is essential for soapmaking)
  - 6 oz of distilled water
  - Optional: essential oils for fragrance (lavender, eucalyptus, etc.)

- **Instructions**:

0. Slowly add the lye to the water (never the other way around) in a well-ventilated area. Stir and let the mixture cool.
1. Melt the coconut oil and combine it with the olive oil.
2. Once both the oil mixture and the lye mixture have cooled to around 100°F, slowly combine them, stirring constantly.
3. Pour the mixture into molds and let it cure for 4–6 weeks before use.

Sustainable Bathing Solutions

Without access to a traditional plumbing system, you'll need to rely on off-grid methods for bathing. The key is to conserve water while ensuring that you stay clean and hygienic.

Solar Shower

A solar shower uses the sun's energy to heat water, allowing you to take a warm shower even without electricity.

- **Materials**: A large, dark-colored container (such as a camping shower bag) filled with water and hung in direct sunlight.
- **Instructions**: Place the container in a sunny spot for a few hours to heat the water, then hang it above head height. Gravity will provide the water pressure needed for a simple, effective shower.

Sponge Bath

In the absence of a shower, sponge baths can help maintain cleanliness while using minimal water.

- **Materials**: A basin of warm water, a washcloth, and biodegradable soap.

- **Instructions**: Wet the washcloth, apply soap, and wipe down your body, focusing on high-sweat areas. Use the water sparingly and avoid wasting any.

14.2 Waste Management and Recycling

Proper waste management is critical in off-grid living to prevent contamination, reduce the risk of disease, and maintain a sanitary living environment. This section covers various methods of handling waste, including human waste, trash, and recyclables.

Building Composting Toilets

A composting toilet is a simple, eco-friendly solution for managing human waste without relying on a traditional septic system or sewage infrastructure. These toilets convert waste into compost, which can later be used to fertilize non-edible plants.

- **Materials**:
    - A large bin or bucket
    - A toilet seat
    - Sawdust or peat moss (to cover waste and control odor)
- **Instructions**:

0. Construct a simple wooden frame to hold the bin or bucket and attach the toilet seat.
1. After each use, cover the waste with sawdust or peat moss to help with decomposition and reduce odor.
2. Once the bin is full, transfer the waste to a dedicated composting area and allow it to break down for at least 6–12 months before using it as fertilizer.

Off-Grid Trash Management and Recycling

Proper disposal of household trash and recyclables is essential for maintaining a clean and sustainable off-grid home. Without regular trash collection services, you'll need to manage waste independently.

Sorting and Reducing Waste

Start by minimizing waste as much as possible. Recycle and repurpose materials, and avoid single-use products whenever possible.

- **Composting Organic Waste**: Food scraps, plant materials, and other organic waste should be composted. Composting reduces the volume of waste while creating valuable fertilizer for your garden.
- **Recycling and Reusing Materials**: Glass, metal, and plastic should be cleaned and stored for future use. For example, glass jars can be reused for food storage, and metal scraps can be repurposed for construction projects.

15: Natural Remedies and First Aid in Off-Grid Living

In an off-grid setting, quick access to medical help may not be available, so it's essential to be prepared with a well-stocked first aid kit and basic knowledge of emergency medical care. In this section, we'll discuss what to include in your off-grid first aid kit, as well as natural remedies and techniques for dealing with common injuries and illnesses.

15.1 Essential First Aid Kit Items for Off-Grid Living

A comprehensive first aid kit is critical for treating injuries and managing health conditions when medical facilities are far away. Here's what you should include:

- **Basic Supplies**: Bandages, gauze, adhesive tape, antiseptic wipes, tweezers, scissors, and a thermometer.
- **Medications**: Pain relievers (such as ibuprofen or acetaminophen), antihistamines, and anti-diarrheal medication.
- **Wound Care**: Sterile wound dressings, antiseptic ointments, and sutures or wound closure strips.
- **Splints and Slings**: For treating sprains or fractures until proper medical help is available.
- **Burn Treatments**: Aloe vera gel and burn ointments for treating burns.

15.2 Herbal Remedies and Natural Health Solutions

In the absence of pharmaceuticals, natural remedies and herbal medicines can be an invaluable resource for maintaining health and treating minor ailments.

- **Echinacea**: Used to boost the immune system and fight off infections like colds and flu.
- **Aloe Vera**: A natural remedy for burns, cuts, and skin irritations.
- **Comfrey**: Known for its healing properties, comfrey can be used in poultices to treat sprains, bruises, and inflammation.
- **Garlic**: A natural antibiotic, garlic can be used to treat infections and boost overall immunity.
- **Lavender**: Known for its calming effects, lavender is useful for treating stress, anxiety, and headaches.

Dealing with Common Off-Grid Medical Issues

When living off the grid, you may face common injuries or illnesses that require immediate attention but aren't severe enough to warrant an emergency evacuation. Knowing how to treat these issues on your own is crucial.

- **Minor Cuts and Scrapes**: Clean the wound with soap and water, apply antiseptic, and cover with a sterile bandage.
- **Burns**: For minor burns, cool the area with running water for several minutes, then apply aloe vera or burn ointment. Cover with a clean dressing.

- **Sprains and Strains**: Rest the injured limb, apply ice to reduce swelling, and use a splint or compression bandage if necessary.
- **Infections**: Clean the infected area thoroughly and apply natural antiseptics like tea tree oil or garlic. For more serious infections, antibiotics from your first aid kit may be required.

Key Takeaways for Off-Grid Hygiene and Health:

- **Personal hygiene** can be maintained using DIY products like homemade toothpaste and soap, and with alternative bathing methods like solar showers or sponge baths.
- **Waste management** is essential for off-grid living. Composting toilets and sustainable trash management systems help reduce contamination and recycle valuable resources.
- **First aid preparedness** and **natural remedies** are crucial for handling medical emergencies and maintaining health when professional care is not readily available.
- **Herbal medicine** offers sustainable, natural solutions for treating common ailments, helping to reduce reliance on pharmaceuticals in an off-grid setting.

By implementing these hygiene and health strategies, you can maintain a safe, clean, and healthy environment while living off the grid, ensuring that you and your family are prepared to handle any challenges that may arise.

16. Home Defense and Security

In an off-grid lifestyle, ensuring the safety and security of your home and family is a top priority. Living in remote areas or without the usual infrastructure means you may be more vulnerable to intruders, wildlife, or other threats. Without access to law enforcement or modern security systems, it's essential to take responsibility for your own home defense. This chapter will guide you through various methods of securing your property, from physical barriers and security systems to personal defense strategies.

16.1 Designing a Perimeter Security System

Securing the perimeter of your property is the first step in creating a comprehensive defense system for your off-grid home. A well-designed perimeter security system deters potential intruders, alerts you to any threats, and provides a clear boundary for your property. In this section, we will explore several off-grid solutions for creating an effective perimeter security system, including solar-powered lighting, alarms, and physical barriers.

Solar-Powered Motion-Activated Floodlights

Lighting is one of the simplest yet most effective deterrents for potential intruders. Solar-powered motion-activated floodlights offer an eco-friendly, sustainable solution that doesn't rely on grid power. These lights provide illumination in key areas when motion is detected, startling intruders and alerting you to their presence.

Advantages of Solar-Powered Floodlights:

- **Energy Independence**: Solar-powered lights harness the energy of the sun, making them ideal for off-grid homes. They don't require an external power source, which means they continue to operate during power outages or in remote areas without access to electricity.
- **Cost-Effective**: Once installed, solar-powered lights are low-maintenance and cost-effective, as they don't require electricity from the grid.
- **Easy Installation**: Solar lights are easy to install and can be placed in any location that receives sunlight, without the need for wiring or electrical expertise.

Where to Install Floodlights:

- **Near Entry Points**: Install floodlights near all entry points to your home, including front and back doors, windows, and garages. This ensures that any movement near these vulnerable areas triggers the lights.
- **Around the Perimeter**: Place floodlights along the perimeter of your property to detect movement before intruders reach your home. Position lights at gates, driveways, and other access points.
- **Near Livestock or Garden Areas**: Protect your livestock and crops by installing floodlights near animal enclosures, barns, and garden plots. This helps deter wildlife or potential thieves.

8.1.2 Setting Up Solar-Powered Alarm Systems

Solar-powered alarms provide an additional layer of security by alerting you when someone or something enters your property. These alarms are triggered by motion or the opening of doors and gates, allowing you to respond to threats before they reach your home.

Types of Solar-Powered Alarms:

- **Basic Motion Detectors**: These alarms are equipped with motion sensors that activate when movement is detected in a designated area. They can emit loud sounds or lights to scare off intruders and alert you.
- **Wireless Systems**: Some solar-powered alarm systems come with wireless features that can send alerts to a mobile device or indoor receiver, keeping you informed even when you're inside or away from the property.
- **Door and Window Sensors**: Install solar-powered sensors on doors, gates, and windows to detect when they are opened. These alarms are ideal for securing entry points and providing early warning if an intruder tries to gain access.

Where to Install Alarms:

- **Perimeter Fencing**: Install motion-activated alarms along your perimeter fencing or at key access points like gates. This provides early warning if someone or something crosses onto your property.
- **Doors and Windows**: Place alarms on all doors and windows, especially those that are easily accessible from the ground. This ensures that any attempt to enter your home will trigger an alert.
- **Driveways and Walkways**: Use motion-activated alarms to monitor driveways or walkways leading to your home. These alarms can be positioned to detect vehicles or people approaching your home, giving you time to prepare or assess the situation.

Fences and Barriers for Physical Security

Physical barriers, such as fences and gates, form the backbone of your perimeter security system. A well-constructed fence defines the boundaries of your property, limits access, and provides a visible deterrent to potential intruders.

Types of Fencing:

- **Wooden Fences**: Solid wooden fences offer privacy and security by blocking the view of your property and making it more difficult for intruders to access. They are ideal for areas close to your home or garden where visibility is a concern.
- **Wire Fences**: Wire fencing, such as barbed wire or welded wire, is cost-effective and suitable for larger areas like livestock enclosures or fields. While it doesn't offer as much privacy, wire fencing is a strong physical barrier that can be difficult to bypass.
- **Electric Fences**: Solar-powered electric fences provide an added layer of security, particularly for protecting livestock or agricultural areas from wildlife or trespassers. The electric current is mild but discourages animals or people from crossing the barrier.

Additional Barrier Options:

- **Hedges and Natural Barriers**: Planting dense hedges or thorny bushes around your property can create a natural barrier that is difficult to pass through. These barriers blend into the landscape while adding an extra layer of security.
- **Gates and Driveway Barriers**: Installing secure gates at entry points such as driveways is essential for controlling who enters your property. Lockable gates with sturdy hinges and latches make it harder for intruders to access your home.

Key Takeaways for Perimeter Security:

- **Solar-powered lighting and alarms** are essential for off-grid security, offering energy-efficient, low-maintenance solutions that keep your property illuminated and alert you to potential threats.

- **Floodlights** should be placed at entry points, along the perimeter, and near high-value areas like livestock enclosures and gardens to detect and deter intruders.
- **Solar-powered alarms** provide early warning by monitoring doors, windows, gates, and driveways for movement or unauthorized entry.
- **Fences and barriers** offer a physical line of defense, preventing intruders from easily accessing your property. Choose from wooden, wire, or electric fences based on your needs and the size of your property.

By implementing these perimeter security measures, you can effectively safeguard your off-grid home, creating a secure boundary that discourages intruders and provides early detection in case of a breach.

16.2 Personal and Family Protection

When living off-grid, personal and family safety is a top priority. Without easy access to emergency services or modern security infrastructure, it's essential to take a proactive approach to protecting yourself, your loved ones, and your property. This section explores practical strategies for personal and family defense, including training and using guard dogs, as well as responsible use of firearms for self-defense.

Guard Dogs: Choosing and Training for Off-Grid Security

A well-trained guard dog is one of the most reliable forms of security for off-grid living. Dogs provide early warnings of potential intruders, protect livestock from predators, and offer companionship. Choosing the right breed and training your dog properly is crucial for ensuring they fulfill their role as protectors while still being safe and well-behaved family members.

1. Best Breeds for Off-Grid Security

Certain dog breeds are naturally suited to guarding and protection tasks. These breeds tend to be loyal, intelligent, and protective, making them ideal for off-grid living.

- **German Shepherd**: Known for their intelligence, loyalty, and versatility, German Shepherds are excellent guard dogs. They are highly trainable and can be taught advanced commands to enhance home security.
- **Great Pyrenees**: This breed excels at livestock protection, guarding animals against predators like coyotes and wolves. They are gentle with family members but fearless when defending their territory.
- **Rottweiler**: Rottweilers are strong and protective, making them a good choice for personal and property defense. They form strong bonds with their owners and are naturally territorial, which helps deter intruders.
- **Belgian Malinois**: Similar to German Shepherds, Belgian Malinois are agile, intelligent, and capable of handling high-stress situations. They require regular exercise and training to channel their energy effectively.

2. Training Your Guard Dog

Proper training is essential for ensuring your dog becomes an effective protector while still being a safe companion for your family. Start training early to build a strong foundation of obedience and security skills.

Basic Obedience Training:

- **Commands**: Teach your dog basic commands like "sit," "stay," "come," and "leave it." These commands are the foundation for more advanced training and ensure you can control your dog in various situations.
- **Recall**: Ensure your dog responds to recall commands (such as "come") immediately, even when distracted by other animals or people. This is crucial for maintaining control in emergency situations.
- **Boundary Training**: Train your dog to patrol the perimeter of your property. Use verbal cues or boundary markers to reinforce where the dog should focus its guarding efforts.

Advanced Guard Training:

- **Barking on Command**: Teach your dog to bark only when necessary—such as when an unfamiliar person or animal approaches the property. This helps avoid unnecessary noise while providing an early alert system.
- **Controlled Aggression**: Train your dog to exhibit controlled aggression. This means they will only act defensively when necessary and will not attack without a clear threat. Use positive reinforcement to reward calm, protective behavior.
- **Socialization**: Guard dogs must be well-socialized to ensure they don't become overly aggressive or fearful. Expose your dog to various people, animals, and environments so they can distinguish between friend and foe.

Firearms for Personal Protection

Firearms can be a critical component of personal and family defense, particularly in remote, off-grid areas where law enforcement may not be readily available. When handled responsibly, firearms provide an effective means of deterring or defending against intruders, both human and animal.

However, it's important to follow safety protocols and ensure that everyone in the household is trained in proper firearm use.

1. Choosing the Right Firearm for Off-Grid Living

The type of firearm you choose depends on your specific security needs, comfort level, and local wildlife. Here are some commonly used firearms for off-grid protection:

- **Shotguns**: Shotguns are versatile weapons suitable for both home defense and hunting. They have a wide spread, making them effective at close range. Shotguns can also be used to deter or eliminate dangerous wildlife like bears or wolves.
- **Rifles**: Rifles offer long-range accuracy and are often used for hunting or protecting livestock from predators. They are effective for defending against threats from a distance and are a staple for off-grid homesteaders in rural areas.
- **Handguns**: Handguns are easy to carry and provide a quick, accessible defense option. They are ideal for personal protection around the home or property. However, they require regular practice to use effectively in high-stress situations.

2. Firearm Safety and Responsible Use

Firearm safety should always be a priority. Before using a firearm for self-defense, ensure that you and your family are thoroughly trained in its operation, safety measures, and storage.

Safety Tips:

- **Secure Storage**: Always store firearms in a locked gun safe, especially if children are present. Only responsible, trained individuals should have access to the firearms. Ensure that ammunition is stored separately.
- **Regular Practice**: Practice regularly to maintain proficiency with your firearm. This includes learning how to load, aim, and fire accurately under pressure. Shooting ranges provide a safe environment for practicing.
- **Trigger Discipline**: Keep your finger off the trigger until you are ready to fire. This prevents accidental discharge and ensures that you only fire in self-defense situations.
- **Proper Identification of Threats**: Before using a firearm, ensure you properly identify the threat. Never shoot unless you are certain the target poses a legitimate danger.

3. Dealing with Wildlife Threats

Off-grid living often involves dealing with wildlife, some of which can pose a danger to your family or livestock. Firearms can be used to deter or eliminate these threats if necessary.

- **Bears**: Bears can be a major threat to livestock, crops, and even people. Use a shotgun or rifle to defend against bear attacks if other deterrents, such as loud noises or bear spray, fail.
- **Coyotes and Wolves**: Predators like coyotes and wolves may target livestock. A rifle is an effective tool for protecting your animals from these predators, especially if they approach your property regularly.
- **Feral Hogs**: In some areas, feral hogs can cause significant damage to crops and infrastructure. Shotguns and rifles are effective for controlling hog populations.

Preparing Your Family for Emergency Situations

Ensuring that your entire family is prepared for emergencies is crucial for maintaining safety. This includes having clear plans, drills, and communication strategies in place.

1. Emergency Drills and Communication Plans

Conduct regular emergency drills to ensure that every family member knows how to react in case of an intruder or other dangerous situation.

- **Evacuation Plans**: Create an evacuation plan with clear routes and safe spots both inside and outside the home. Practice these plans regularly so everyone knows how to quickly and calmly escape.
- **Communication Protocols**: Establish clear communication protocols, especially for large properties or if family members often work away from the main house. Use radios or other communication devices to stay in touch.
- **Code Words**: Develop family code words for different situations, such as needing help, evacuating, or indicating that it's safe to return.

2. Teaching Children About Safety

Children should be educated about personal safety and how to react in emergencies, but without instilling unnecessary fear. Age-appropriate lessons about strangers, wildlife, and what to do in dangerous situations are important.

- **Stranger Awareness**: Teach children to recognize when someone is a stranger and how to react if approached by unfamiliar people. Emphasize the importance of staying near the home or a trusted adult.
- **Safe Zones**: Designate safe zones on your property where children can go in case of danger. These should be areas that are easy to access and well-protected from threats.
- **Firearm Safety**: If firearms are in the home, teach children about the dangers of handling guns. Make sure they understand that guns are not toys and that they should never handle them without adult supervision.

Key Takeaways for Personal and Family Protection:

- **Guard dogs** are an excellent first line of defense for off-grid homes. Choose the right breed and train your dog to protect your property while remaining well-behaved with family members and guests.
- **Firearms** provide an effective means of defense, but safety must always come first. Ensure that everyone in your household is trained in the proper use and storage of firearms.
- **Wildlife threats** are a reality for many off-grid homes. Be prepared to defend against dangerous animals like bears, coyotes, and feral hogs, but prioritize non-lethal deterrents when possible.
- **Family preparedness** is essential. Conduct regular emergency drills, establish communication protocols, and ensure that everyone, including children, knows how to react in dangerous situations.

By implementing these strategies, you can ensure the safety and protection of your family in an off-grid environment, whether from human intruders or wildlife threats. Proper preparation and training are key to creating a secure and resilient homestead.

16.3 Fortifying Your Home

In an off-grid environment, your home serves as both a shelter and a fortress. Without the safety net of modern infrastructure, it's essential to take extra steps to fortify your home against potential threats. This section covers methods to reinforce your doors and windows, create secure living spaces, and build safe rooms or shelters. These strategies will help protect your home from intruders, wildlife, and natural disasters, ensuring you and your family remain safe in any situation.

Reinforcing Doors and Windows

The most common entry points for intruders are doors and windows, making them critical areas to reinforce. Strengthening these vulnerable points can prevent break-ins and give you valuable time to react if someone attempts to enter your home.

1. Solid Core or Steel Doors

One of the first steps in home fortification is upgrading your doors. Hollow core doors, typically used for interior spaces, are easy to break through, while solid core or steel doors offer much more protection.

- **Solid Core Doors**: These doors are made from dense wood or composite materials that are harder to force open. Install solid core doors at all exterior entry points to make it more difficult for intruders to break through.
- **Steel Doors**: Steel doors provide an even higher level of protection. They are durable, fire-resistant, and can be reinforced with additional layers of security, such as deadbolts or steel door frames.

2. Reinforced Door Frames and Locks

Even the strongest door is only as secure as its frame and lock. Reinforcing these elements is essential to prevent forced entry.

- **Steel Door Frames**: Replace wooden door frames with steel frames to make them harder to break or split. This is especially important for off-grid homes in remote areas where help may be far away.
- **Deadbolts and Locking Mechanisms**: Install heavy-duty deadbolts with long screws that penetrate deep into the door frame. A double-cylinder deadbolt, which requires a key to unlock from both sides, adds an extra layer of security.
- **Strike Plates**: Reinforce the strike plate (the metal plate where the deadbolt enters the door frame) by using longer screws and upgrading to a high-security plate. This helps prevent the door from being kicked in.

3. Window Security

Windows are another vulnerable point of entry, especially ground-level windows or those that are easily accessible from the outside. Strengthening your windows with barriers, glass treatments, and secondary locks can make it harder for intruders to gain access.

- **Security Bars**: Installing metal bars over windows is a highly effective way to prevent break-ins. These bars can be placed on the inside or outside of the window frame. For emergency purposes, use bars with quick-release mechanisms that allow you to exit the home if necessary.
- **Shatter-Resistant Films**: Apply shatter-resistant window films to prevent glass from breaking easily. These films hold the glass together, even when impacted, making it harder for intruders to enter through broken windows.
- **Secondary Window Locks**: Install secondary locks on all windows, especially sliding or double-hung windows. These locks prevent windows from being forced open from the outside.

4. Installing Security Shutters

Security shutters provide an additional layer of protection for windows and doors. They can be closed and locked to seal off vulnerable entry points, protecting your home not only from intruders but also from severe weather events such as storms or hurricanes.

- **Metal or Wooden Shutters**: Choose sturdy materials like steel, aluminum, or reinforced wood for maximum durability. Shutters can be hinged or roll-up, allowing you to open and close them as needed.
- **Storm Shutters**: If you live in an area prone to extreme weather, storm shutters can double as security and weather protection, shielding your windows from both intruders and natural disasters.

Creating Safe Rooms or Shelters

A safe room provides a secure place to retreat in the event of an intruder, natural disaster, or other emergency. Safe rooms are designed to withstand forced entry and protect occupants from danger, offering you time to call for help or wait for the threat to pass.

1. Choosing a Safe Room Location

The ideal safe room should be located in the most secure part of your home, away from easily accessible areas like exterior walls or windows. Consider the following locations for a safe room:

- **Basement or Cellar**: Underground areas, such as basements or cellars, provide natural protection against intruders and are often well-insulated against extreme weather conditions.
- **Interior Rooms**: If you don't have a basement, choose an interior room with no windows, such as a large closet, pantry, or bathroom. Interior rooms are more difficult for intruders to reach and provide added layers of security.

2. Reinforcing Your Safe Room

Safe rooms need to be built or retrofitted to resist forced entry and provide long-term protection. Here's how you can fortify your safe room:

- **Reinforced Walls**: Add steel reinforcement to the walls of the safe room or use concrete blocks to strengthen them. Reinforced walls can resist bullets, forced entry, and fire.
- **Steel Doors**: Install a steel door with a heavy-duty locking mechanism to secure the room. The door should be solid and designed to resist break-ins. Ensure the door opens inward so it cannot be easily forced open.
- **Ventilation and Communication**: Install a small, secure ventilation system to provide fresh air if you need to stay in the room for an extended period. Keep a communication device like a phone, radio, or satellite communicator inside the safe room to call for help if needed.

3. Stocking Your Safe Room

In case of an extended emergency, your safe room should be stocked with basic survival supplies. These supplies will allow you to wait out a dangerous situation in safety.

- **Food and Water**: Store enough non-perishable food and bottled water to last at least 72 hours. Ensure that the water is easily accessible and enough for everyone who might be in the safe room.
- **First Aid Kit**: A fully stocked first aid kit is essential for treating minor injuries while you're inside the safe room.
- **Tools and Weapons**: Keep basic tools, such as a crowbar or wrench, in the safe room in case you need to break free from debris or blocked exits. If necessary, store a firearm or non-lethal weapon in the room for personal protection.
- **Lighting**: Keep a flashlight or battery-powered lantern on hand to ensure you have light in case of a power outage.
- **Communication Devices**: Ensure you have a means of communication, such as a charged cell phone, walkie-talkie, or emergency radio, to stay in touch with authorities or family members outside the room.

Key Takeaways for Fortifying Your Home:

- **Reinforcing doors and windows** is essential for preventing break-ins and ensuring the security of your home. Install solid core or steel doors, reinforce door frames, and add window bars or shatter-resistant films.
- **Security shutters** provide an additional layer of protection for your windows and doors, shielding your home from intruders and extreme weather conditions.
- **Safe rooms** are a crucial part of your home defense strategy, offering a secure retreat in case of emergency. Reinforce walls and doors, and stock the room with essential supplies like food, water, and communication devices.
- **Preparation is key**: Regularly assess your home's vulnerabilities and reinforce any weak points to ensure you are ready for both natural disasters and potential security threats.

By fortifying your home and creating a secure environment, you can protect your family from intruders and natural disasters, ensuring that you are well-prepared for any situation. Proper planning and reinforcement will give you peace of mind and help keep your off-grid home safe.

17. Long-Term Planning and Resilience

Off-grid living isn't just about short-term survival; it's about creating a sustainable, resilient lifestyle that allows you to thrive over the long term. Chapter 9 explores long-term projects and strategies that will help you achieve true self-sufficiency, from advanced resource management techniques to mental and emotional resilience. These projects require planning and commitment but will ultimately provide the security, independence, and sustainability you need to live comfortably off the grid for years to come.

Building long-term self-sufficiency involves careful planning and forethought. While short-term solutions are essential for immediate survival, long-term planning focuses on creating a sustainable system that can support you and your family for years without relying on external resources.

17.1 Assessing and Managing Resources

To live sustainably off-grid, you need to assess the availability of your resources and manage them efficiently to ensure that they are renewable and sufficient to meet your needs.

- **Water Resources**: Secure a renewable water source that can meet your long-term needs. This might involve drilling a well, setting up large-scale rainwater collection systems, or maintaining a nearby natural water source. Ensure that you have a reliable method for purifying and storing water, and account for changes in climate that may affect water availability.
- **Energy Independence**: Whether you rely on solar, wind, or hydroelectric power, maintaining a renewable energy system is key to long-term off-grid living. Assess your energy needs carefully, considering future growth, and make sure your system is scalable to handle increased demand.
- **Food Production**: Plan for year-round food production through gardening, livestock raising, foraging, and hunting. Consider methods of preserving food long-term, such as canning, drying, or fermenting, to ensure you have enough food through the winter months or in case of crop failure.

Building a Resilient Off-Grid System

Creating a resilient off-grid ecosystem is about more than just surviving—it's about thriving independently and sustainably, no matter the challenges. A resilient ecosystem integrates food production, energy generation, water management, and waste disposal into a cohesive, sustainable whole. By focusing on long-term resilience, you can build a homestead that not only provides for your needs but also regenerates the land and resources around you. This section explores strategies to build and maintain a resilient off-grid ecosystem that will sustain you for the long haul.

---

1. Integrating Key Systems: Energy, Water, Food, and Waste

To build resilience, you need to interconnect the essential systems that support your off-grid lifestyle. Integrating energy, water, food, and waste systems ensures that resources flow efficiently, with minimal waste and maximum sustainability.

- **Energy**: Your energy system should be diverse and scalable to handle fluctuations in power generation, especially during times of reduced sunlight or wind. A resilient energy system might include:
  - **Solar Power**: Solar panels are often the backbone of off-grid energy, but they need to be supplemented during low-light periods.
  - **Wind Power**: Installing small wind turbines can complement solar by generating power when it's windy, especially at night or during cloudy days.
  - **Battery Storage**: A robust battery bank is essential for storing energy. Consider adding **redundant systems**, such as multiple battery banks or a backup generator, to ensure you always have a power source.
- **Water**: Water resilience involves capturing, storing, and recycling water to ensure a continuous supply, even during droughts or extreme weather.
  - **Rainwater Harvesting**: Collect rainwater from rooftops and store it in cisterns or tanks for irrigation, drinking, and household use.
  - **Greywater Recycling**: Reuse water from sinks, showers, and washing machines for irrigation. By filtering and recycling greywater, you reduce overall water consumption and ensure that every drop is used efficiently.
  - **Ponds and Natural Water Sources**: If you have access to a pond, river, or well, integrate these into your water system, ensuring you have a reliable source of fresh water.
- **Food**: A resilient food system is diverse, incorporating multiple sources of nutrition and food production techniques.
  - **Gardening**: Build a survival garden using permaculture principles (see Chapter 6) to grow a variety of crops year-round.
  - **Livestock**: Raising chickens, goats, pigs, or cattle provides meat, milk, eggs, and manure for composting. Livestock should be part of an integrated system that includes rotational grazing and manure management to maintain soil fertility.
  - **Aquaculture**: Raising fish in ponds or tanks can provide an additional source of protein and complement your gardening efforts by creating a closed-loop system (fish waste can fertilize crops).
- **Waste Management**: Waste should be seen as a resource in a resilient ecosystem. Implement systems to recycle organic waste into valuable resources.
  - **Composting**: Composting kitchen scraps, garden waste, and animal manure creates nutrient-rich compost that can be used to improve soil health.

  - **Humanure**: A **composting toilet** system can turn human waste into safe, usable compost, reducing the need for water-based toilets and minimizing environmental impact.
  - **Biogas**: Organic waste, such as manure and food scraps, can be used in a biogas digester to produce methane, which can be used for cooking or heating.

---

2. Creating Wildlife Corridors: Supporting Local Biodiversity

Building a resilient off-grid ecosystem also means supporting local biodiversity. Wildlife corridors allow native animals, insects, and pollinators to move freely through your land, helping to maintain a balanced ecosystem.

- **Pollinator Habitats**: Create pollinator-friendly spaces by planting wildflowers, shrubs, and fruit trees that attract bees, butterflies, and other pollinators. These insects are essential for the pollination of crops, ensuring a steady food supply.
- **Water Sources for Wildlife**: Include small ponds or water catchment areas to provide drinking water for birds, amphibians, and small mammals. These water sources also help regulate insect populations by attracting natural predators.
- **Shelter and Habitat**: Leave sections of your land undisturbed to provide shelter for wildlife. For example, leave dead logs, brush piles, and dense undergrowth to provide nesting sites for birds, reptiles, and insects.

By encouraging wildlife on your property, you help control pests, improve soil health, and ensure that your ecosystem remains balanced and resilient.

---

3. Soil Regeneration and Management

Healthy soil is the foundation of a resilient ecosystem. By continually improving soil quality, you ensure long-term productivity and fertility in your garden and pastures. Soil regeneration techniques focus on maintaining and increasing organic matter, improving water retention, and preventing erosion.

- **Composting and Mulching**: Adding compost to your soil builds organic matter, which improves water retention and nutrient availability. Use organic mulches like straw, leaves, or wood chips to suppress weeds, retain moisture, and break down over time to enrich the soil.
- **Cover Cropping**: Plant cover crops like clover, rye, or alfalfa during the off-season to prevent soil erosion, fix nitrogen, and build organic matter. Cover crops protect the soil from the elements and add nutrients back into the soil as they decompose.
- **No-Till Gardening**: Avoiding tillage minimizes soil disruption and helps maintain a healthy soil structure. No-till gardening preserves the beneficial microorganisms in the soil and reduces erosion.
- **Rotational Grazing**: For those raising livestock, rotational grazing helps maintain pasture health by allowing grass to regrow before animals return to graze. This method prevents overgrazing, reduces soil compaction, and encourages deep-rooted plants that stabilize the soil.

---

4. Water Management for Resilience

In an off-grid ecosystem, efficient water management is crucial for long-term survival, especially in regions prone to drought or erratic weather patterns. Incorporating multiple methods for capturing, storing, and using water ensures that you remain resilient to water shortages.

- **Building Swales**: Swales are shallow ditches built on contour lines that capture rainwater and allow it to infiltrate into the soil. This method reduces runoff, conserves water, and improves soil moisture levels, making it an excellent strategy for dry climates.
- **Rainwater Catchment Systems**: Design a system to capture rainwater from rooftops and store it in large tanks or cisterns. The water can then be used for irrigation, drinking, or household needs. Consider building multiple storage tanks to hold enough water for extended dry periods.
- **Greywater Systems**: Install greywater systems to recycle water from household uses (like washing dishes and showers) and use it to irrigate your garden. This reduces your total water consumption and ensures that water is used more than once.

---

5. Building Redundancy and Flexibility into Your Systems

A resilient off-grid ecosystem is one that can adapt to challenges and unexpected events. By building redundancy and flexibility into your systems, you can prepare for the unexpected and ensure that your essential needs—food, water, and energy—are always met.

- **Multiple Energy Sources**: Relying solely on one energy source (such as solar) can leave you vulnerable if the weather doesn't cooperate. Incorporate multiple energy sources, like wind turbines or backup generators, to provide a safety net during energy shortages.
- **Backup Water Supplies**: Even with a rainwater collection system, it's wise to have backup sources of water, such as a pond, well, or water stored in containers. This redundancy ensures that you won't run out of water during dry seasons or emergencies.
- **Diversify Food Production**: Growing multiple types of crops and raising different kinds of livestock increases food security. If one crop fails or a disease affects a particular type of livestock, you still have other food sources to rely on.
- **Emergency Preparedness**: Have emergency plans and backup systems in place. For example, keep extra fuel for generators, store long-term food supplies, and create a first-aid station. Regularly test and update these systems to ensure they're ready when needed.

Key Takeaways for Building a Resilient Off-Grid Ecosystem:

- **Integrate Essential Systems**: Energy, water, food, and waste systems should be interconnected to maximize efficiency and sustainability.
- **Support Biodiversity**: Encourage local wildlife and pollinators to strengthen the natural balance of your ecosystem.
- **Regenerate Soil**: Focus on building soil health through composting, mulching, cover cropping, and no-till methods to maintain fertility and productivity.

- **Manage Water Wisely**: Use rainwater harvesting, greywater systems, and swales to conserve water and build resilience against droughts.
- **Build Redundancy**: Ensure you have backup systems for energy, water, and food to stay prepared for unexpected challenges.

By focusing on these strategies, you can build a resilient, sustainable off-grid ecosystem that will thrive for years, providing security and self-sufficiency even in the face of changing conditions or unforeseen challenges.

17.2 Mental and Emotional Resilience

Long-term self-sufficiency isn't just about managing resources—it's also about maintaining your mental and emotional well-being. Living off the grid can be isolating, stressful, and physically demanding, so building mental resilience is just as important as securing food and energy.

Managing Stress and Isolation

Isolation, particularly in remote areas, can be a major challenge in off-grid living. Without the social networks and conveniences of modern life, feelings of loneliness and stress can build up, making it difficult to stay focused and productive.

- **Stay Connected**: Maintain regular contact with friends, family, and other off-grid communities through radios, satellite phones, or occasional trips into town. Connecting with others helps alleviate isolation and provides support during challenging times.
- **Create Routines**: Develop daily and weekly routines to structure your life. A consistent schedule helps manage stress and ensures you're using your time effectively to meet your needs. Routines also make long-term projects more manageable and help you avoid burnout.
- **Mindfulness Practices**: Incorporate mindfulness or meditation practices into your daily routine. These techniques can help reduce stress, improve mental clarity, and boost emotional resilience. Spending time in nature, away from technology, is a natural benefit of off-grid living and can help you stay grounded.

17.2 Developing a Resilient Mindset

Off-grid living often comes with unexpected challenges, from bad weather to crop failures to mechanical breakdowns. Building a resilient mindset will help you face these challenges head-on and remain adaptable and resourceful.

- **Problem-Solving Skills**: One of the most valuable traits for long-term off-grid success is the ability to solve problems as they arise. Learn how to fix and maintain your equipment, troubleshoot energy or water systems, and adapt when resources are scarce.
- **Flexibility and Adaptability**: Accept that things won't always go according to plan. The weather may not cooperate, or unexpected events may force you to change course. Being adaptable and flexible ensures that you're prepared for whatever comes your way.

- **Set Achievable Goals**: While long-term self-sufficiency can feel overwhelming, breaking down larger projects into smaller, achievable goals can help you stay motivated. Celebrate each success and continue to build on your accomplishments over time.

---

18. Expanding and Improving Your Homestead

As you become more established in your off-grid lifestyle, you may want to expand your homestead or improve existing systems to increase comfort, productivity, and sustainability. This could involve scaling up food production, expanding living spaces, or upgrading energy systems.

Scaling Up Food Production

Once you've mastered the basics of gardening and livestock management, you can scale up your food production to increase self-sufficiency or even trade with nearby communities.

- **Greenhouses and Cold Frames**: Building greenhouses or cold frames allows you to grow food year-round, regardless of the weather. This is especially important in areas with short growing seasons or harsh winters.
- **Aquaponics Systems**: Aquaponics combines fish farming and hydroponic gardening, creating a sustainable, water-efficient system for growing both fish and vegetables. These systems can be highly productive, especially in areas with limited water or arable land.
- **Expanding Livestock Operations**: If you have the space and resources, expanding your livestock operation can provide a more reliable source of protein and other animal products. Raising additional animals, such as goats, pigs, or bees, will increase your food security and provide more resources for trading or bartering.

Enhancing Energy and Water Systems

As your off-grid homestead grows, you may need to upgrade or expand your energy and water systems to meet increased demand.

- **Solar Panel Expansion**: If your household grows or your energy needs increase, consider expanding your solar power system by adding additional panels or upgrading your battery storage. This ensures that your energy supply remains consistent, even as your needs change.
- **Water Conservation and Reuse**: Install greywater recycling systems to reuse water from sinks, showers, and laundry for irrigation or other non-potable uses. This reduces your overall water consumption and ensures a more sustainable water supply in the long term.

Adding Comfort and Convenience

Living off the grid doesn't have to mean sacrificing comfort. Once your basic needs are met, you can begin adding improvements to make your home more comfortable and efficient.

- **Heating and Cooling Systems**: Install efficient wood stoves or solar heating systems to keep your home warm in the winter. For cooling, consider adding passive cooling techniques such as strategically placed windows or installing a solar-powered fan system.
- **Waste Management Upgrades**: Consider building a more permanent composting toilet system or upgrading to a biogas generator that can turn human waste into usable energy.

Key Takeaways for Long-Term Projects for Self-Sufficiency:

- **Resource management** is key to long-term sustainability. Focus on renewable energy, water conservation, and year-round food production to ensure a stable supply of essential resources.
- **Building resilience**—both mental and emotional—is critical for thriving in an off-grid lifestyle. Manage stress, adapt to challenges, and maintain a problem-solving mindset.
- **Scaling up your homestead** through expanded food production, enhanced energy systems, and added comfort features can make off-grid living more sustainable and enjoyable.
- **Flexibility and planning** are crucial to your long-term success. Be prepared to adapt, improve your systems over time, and continuously work toward greater self-sufficiency.

By focusing on these long-term strategies and projects, you can transform your off-grid lifestyle from basic survival to a thriving, sustainable way of life. With the right planning, mindset, and resources, you can achieve true independence and self-reliance.

19. Cultivating Medicinal Herbs and Wellness

Living off-grid requires more than just meeting basic needs like food, water, and shelter; it's equally important to take care of your **physical and mental health**. In an environment where access to modern healthcare and medication may be limited, it's crucial to know how to maintain your family's well-being using natural methods. This chapter will guide you through cultivating **medicinal herbs**, preparing **natural remedies**, creating an **off-grid first-aid kit**, and strategies for maintaining **daily health** and a **resilient mindset**.

19.1. Growing and Using Medicinal Herbs

Medicinal herbs are one of the most valuable resources for off-grid living, providing a sustainable way to care for your family's health without relying on pharmaceutical products. When cultivated and prepared correctly, these herbs can offer natural remedies for everything from common colds to skin irritations, digestive issues, and more. In this section, we'll dive into how to cultivate, harvest, and use some of the most beneficial medicinal herbs for an off-grid lifestyle.

Echinacea (Echinacea purpurea)

Echinacea is widely known for its immune-boosting properties and is often used to ward off colds, infections, and even speed up recovery time. It's a powerful herb to have in your off-grid medicinal garden due to its multiple uses and relatively easy cultivation.

- **How to Grow**:
  Echinacea thrives in full sunlight and well-drained soil. It can tolerate drought once established, making it an excellent herb for areas with low water access. Sow the seeds directly in your garden in the spring after the last frost. The plant typically takes one to two years to fully mature, but it requires minimal care once established.
  - **Planting**: Space seeds 12-18 inches apart, as the plant will spread and form clumps.
  - **Harvesting**: Harvest the flowers during peak bloom (late summer to early fall). You can also harvest the roots of mature plants for medicinal use in the fall.
- **Medicinal Uses**:
  - **Immune support**: Echinacea boosts the immune system and helps reduce the duration of colds and infections.
  - **How to Use**: Brew dried echinacea flowers as a tea. For a stronger effect, create a tincture by steeping the roots or flowers in alcohol for several weeks, then strain and store in a dark glass bottle.
  - **Wound treatment**: The anti-inflammatory and antimicrobial properties of echinacea make it an excellent addition to salves for cuts and wounds.

Calendula (Calendula officinalis)

Calendula, also known as pot marigold, is prized for its anti-inflammatory and healing properties. It is commonly used in skincare products for treating wounds, rashes, and irritations.

- **How to Grow**:
  Calendula is a hardy plant that thrives in most climates. It prefers full sun but will tolerate partial shade. Plant seeds directly in the garden in the spring after the last frost. It grows well in well-drained soil but is quite forgiving if the soil is less than perfect.
  - **Planting**: Sow the seeds about 6-12 inches apart, as they can spread out and self-seed.
  - **Harvesting**: Harvest the flowers throughout the summer as they bloom. The more you pick the flowers, the more the plant will produce.
- **Medicinal Uses**:
  - **Skin healing**: Calendula has powerful skin-healing properties. It can be used to treat cuts, scrapes, burns, and rashes. Calendula-infused oils and salves are commonly applied to the skin to promote healing.
  - **How to Use**: To make a calendula salve, infuse dried calendula flowers in oil (such as olive or coconut oil) for several weeks. Strain the oil and mix it with melted beeswax to create a healing salve.
  - **Anti-inflammatory**: Calendula can also be used internally as a tea to help with inflammatory conditions such as digestive issues or sore throats.

Peppermint (Mentha piperita)

Peppermint is a versatile herb with many medicinal uses, including relief from headaches, digestive issues, and muscle pain. It's easy to grow and offers both culinary and medicinal benefits.

- **How to Grow**:
  Peppermint prefers partial shade and moist, well-drained soil. However, it can grow aggressively, so it's recommended to plant it in containers to control its spread. Peppermint is perennial, meaning it will come back year after year.
  - **Planting**: Peppermint can be started from seeds, cuttings, or transplants. Space plants about 18 inches apart and keep the soil consistently moist.
  - **Harvesting**: Harvest peppermint leaves throughout the growing season. The best time to pick the leaves is in the morning after the dew has dried but before the sun becomes too strong.
- **Medicinal Uses**:
  - **Digestive aid**: Peppermint is well known for its ability to relieve digestive discomfort. It can help with bloating, gas, and nausea.
  - **How to Use**: Brew fresh or dried peppermint leaves in hot water to make a soothing tea for digestive issues. The essential oil of peppermint can also be applied topically to help with nausea.
  - **Headache relief**: The cooling sensation of peppermint essential oil can help relieve headaches and migraines.
  - **How to Use**: Dilute peppermint essential oil in a carrier oil (such as coconut oil) and apply it to the temples or back of the neck for quick relief.

Lavender (Lavandula angustifolia)

Lavender is renowned for its calming properties and is commonly used in aromatherapy to reduce stress and promote relaxation. It's also a natural antiseptic, making it useful for treating cuts and burns.

- **How to Grow**:
  Lavender thrives in full sun and well-drained soil. It does best in areas with warm, dry climates. Plant lavender seeds or transplants in the spring, and be sure to give the plants plenty of space to grow, as they can become large and bushy.
  - **Planting**: Space lavender plants about 2-3 feet apart, as they will grow to a large size. Lavender is drought-tolerant once established but needs regular watering in its early stages.
  - **Harvesting**: Harvest lavender flowers when they are just starting to open. This is when the essential oil concentration is highest. Cut the stems in the morning and hang them upside down to dry in a cool, dark place.
- **Medicinal Uses**:
  - **Calming and stress relief**: Lavender is commonly used to reduce anxiety, improve sleep, and relieve stress.
  - **How to Use**: Add a few drops of lavender essential oil to a diffuser to promote relaxation, or place dried lavender flowers in a sachet near your pillow for a calming effect at night.
  - **Wound healing and antiseptic**: Lavender's antiseptic properties make it useful for treating cuts and burns.
  - **How to Use**: Dilute lavender essential oil in a carrier oil and apply it to burns, cuts, or insect bites to promote healing and reduce pain.

Yarrow (Achillea millefolium)

Yarrow is an excellent herb to have in your off-grid medicinal garden because of its wound-healing abilities and its usefulness in treating fevers and colds.

- **How to Grow**:
  Yarrow is a hardy perennial that can grow in a variety of conditions, from full sun to partial shade and in poor, dry soil. It's an easy plant to maintain and will spread quickly.
  - **Planting**: Sow yarrow seeds directly in the garden in the spring or fall. The plants should be spaced about 12-18 inches apart. Yarrow requires little care once established and is drought-tolerant.
  - **Harvesting**: Harvest yarrow flowers and leaves when they are in full bloom, typically in mid-summer. Hang the stems upside down to dry in a cool, dark place.
- **Medicinal Uses**:
  - **Wound treatment**: Yarrow is a natural astringent and can help stop bleeding and promote clotting.
  - **How to Use**: Make a poultice from fresh yarrow leaves to place on cuts and wounds, or brew a tea from dried yarrow flowers for internal use to reduce fever and fight colds.
  - **Fever reduction**: Yarrow tea can help bring down fevers by promoting sweating.

19.2. Making DIY Natural Remedies

Having medicinal herbs in your garden is only the first step in achieving natural, off-grid healthcare. To make the most of these plants, you need to learn how to transform them into usable **remedies** such as **salves**, **tinctures**, **infusions**, and **essential oils**. In this section, we will explore how to create various DIY remedies that can be used for common health issues such as cuts, burns, colds, headaches, digestive discomfort, and more. These remedies are easy to make and provide a natural, sustainable way to take care of your family's health.

Creating Herbal Salves

Herbal salves are soothing ointments made by infusing herbs into oils and combining them with beeswax. They are ideal for treating skin conditions like cuts, burns, rashes, and insect bites. Here's how you can make a basic herbal salve.

- **Ingredients Needed**:
    - **Dried herbs** (such as calendula, plantain, or comfrey).
    - **Carrier oil** (such as olive oil, coconut oil, or almond oil).
    - **Beeswax** (acts as a thickening agent).
    - **Essential oils** (optional for added benefits and fragrance).
    - **Jars or tins** for storing the salve.
- **Basic Process for Making a Salve**:

    0. **Infusing the oil**: First, choose the herbs you want to use based on their healing properties. For example, calendula is excellent for skin healing, while comfrey is good for inflammation. Place about 1 cup of dried herbs into a jar and cover them completely with 1 to 2 cups of carrier oil. Let the herbs infuse for 3 to 6 weeks in a warm, sunny spot, shaking the jar every few days. Alternatively, you can speed up the process by heating the jar in a double boiler over low heat for 3 to 5 hours.
    1. **Strain the oil**: After the infusion is complete, strain the herbs out using a fine mesh strainer or cheesecloth, making sure to squeeze out as much oil as possible.
    2. **Melt the beeswax**: In a double boiler, melt about 1 ounce of beeswax for every 4 ounces of infused oil. Stir the mixture until the beeswax is fully melted.
    3. **Combine and cool**: Remove the oil and beeswax mixture from heat and add a few drops of essential oils if desired (lavender essential oil is a great choice for added healing properties). Pour the mixture into tins or jars and let it cool and solidify. Your salve is now ready for use!

- **Uses for Herbal Salves**:
    - Apply calendula salve to cuts, scrapes, and burns to speed healing and soothe irritation.
    - Use comfrey salve for muscle aches, joint pain, or inflammation.
    - Lavender-infused salve can be used for minor burns, rashes, or insect bites to reduce pain and inflammation.

Making Herbal Tinctures

Tinctures are concentrated liquid extracts made by steeping herbs in alcohol or glycerin. They are a powerful way to administer herbal remedies and can be stored for long periods, making them a valuable addition to your off-grid medicine cabinet.

- **Ingredients Needed**:
    - **Dried or fresh herbs** (echinacea, peppermint, or elderberry are great choices).
    - **Alcohol** (such as vodka or brandy, at least 40% alcohol content).
    - **Glass jar** with a tight lid.
    - **Dropper bottles** for storage.

- **Basic Process for Making a Tincture**:

    0. **Prepare the herbs**: Finely chop fresh herbs or use dried herbs (use about 1 cup of fresh herbs or 1/2 cup of dried herbs for a pint of alcohol).
    1. **Add alcohol**: Place the herbs in a glass jar and pour enough alcohol over them to completely cover the herbs by at least an inch. Make sure the herbs are fully submerged.
    2. **Steeping process**: Seal the jar tightly and place it in a cool, dark place for 4 to 6 weeks, shaking the jar daily to help with the extraction process.
    3. **Strain and store**: After 4 to 6 weeks, strain the tincture through a fine mesh strainer or cheesecloth and pour the liquid into dropper bottles. Label the bottles with the herb name and date.

- **Uses for Herbal Tinctures**:
    - **Echinacea tincture**: Take a few drops diluted in water at the first sign of a cold or infection to boost the immune system.
    - **Peppermint tincture**: Use for digestive issues such as bloating or nausea. Add a few drops to a glass of water.
    - **Elderberry tincture**: A potent remedy for colds and flu, take daily during the cold season to strengthen your immune system.

Herbal Infusions and Teas

Herbal infusions and teas are one of the easiest and most effective ways to consume medicinal herbs. By steeping herbs in hot water, you can extract their healing properties and create a soothing drink to treat a variety of ailments.

- **Ingredients Needed**:
    - **Dried or fresh herbs** (such as chamomile, peppermint, or ginger).
    - **Hot water**.
    - **Teapot or infuser**.
- **Basic Process for Making an Herbal Tea**:

    0. **Prepare the herbs**: Use about 1 tablespoon of dried herbs or 2 tablespoons of fresh herbs per cup of water.
    1. **Boil water**: Bring water to a boil and pour it over the herbs in a teapot or infuser.
    2. **Steep the herbs**: Let the herbs steep for 5 to 15 minutes, depending on the herb and the desired strength of the tea.
    3. **Strain and drink**: Strain the herbs out and enjoy your tea. Add honey or lemon for extra flavor and medicinal benefits.

- **Uses for Herbal Teas**:
    - **Chamomile tea**: A gentle tea that helps with relaxation, insomnia, and digestive issues.
    - **Peppermint tea**: Great for soothing an upset stomach, relieving gas, and reducing nausea.
    - **Ginger tea**: A warming tea that aids digestion, reduces inflammation, and helps with cold symptoms.

Making Essential Oils

Essential oils are highly concentrated plant extracts that can be used for a variety of therapeutic purposes. While distilling essential oils at home requires specialized equipment, there are simpler ways to create infused oils that can be used in homemade remedies.

- **Ingredients Needed**:
  - **Fresh herbs** (lavender, peppermint, or rosemary are ideal for infused oils).
  - **Carrier oil** (such as olive oil or sweet almond oil).
  - **Glass jar**.
- **Basic Process for Making an Infused Oil**:

  0. **Prepare the herbs**: Lightly crush the fresh herbs to release their oils.
  1. **Add oil**: Place the herbs in a glass jar and cover them with carrier oil, making sure the herbs are fully submerged.
  2. **Infusion process**: Seal the jar and place it in a sunny window for 1 to 2 weeks, shaking it daily to help release the essential oils.
  3. **Strain and store**: Strain the herbs from the oil and pour the infused oil into a dark glass bottle for storage.

- **Uses for Infused Oils**:
  - **Lavender oil**: Apply to burns, cuts, or insect bites for quick healing and pain relief.
  - **Peppermint oil**: Massage into the temples to relieve headaches or add a few drops to a bath for muscle relaxation.
  - **Rosemary oil**: Rub into sore muscles or use as a hair treatment to stimulate hair growth.

Compresses and Poultices

Herbal compresses and poultices are excellent for treating localized pain, inflammation, or injuries. These remedies involve applying herbs directly to the skin, either in the form of a hot compress or a moist poultice.

- **Hot Compress**:
  - **Ingredients Needed**: Cloth or towel, hot water, herbs (such as chamomile or arnica).
  - **How to Use**: Soak the herbs in hot water, then dip a cloth or towel in the liquid. Wring out the excess water and apply the compress to the affected area for 15-20 minutes. This is great for muscle aches, sprains, or bruises.
- **Poultice**:
  - **Ingredients Needed**: Fresh herbs (such as plantain, yarrow, or comfrey), mortar and pestle.
  - **How to Use**: Crush fresh herbs into a paste using a mortar and pestle. Apply the paste directly to wounds, insect bites, or inflamed skin, and cover with a cloth. Leave the poultice in place for several hours, changing as needed.
- **Uses for Compresses and Poultices**:
  - **Arnica compress**: For soothing sore muscles and reducing bruising.
  - **Plantain poultice**: Effective for drawing out toxins from insect bites, stings, and wounds.
  - **Comfrey poultice**: Speeds up the healing of bone fractures, sprains,

20. Embracing the Journey to Self-Sufficiency

20.1 Final Reflections on Long-Term Off-Grid Living

When living off-grid, being prepared for medical emergencies is essential, especially since access to professional medical care may be delayed or unavailable. A well-stocked **off-grid first-aid kit** can be a lifesaver, allowing you to treat a wide variety of injuries and ailments with both **conventional supplies** and **natural remedies**. This section will guide you through assembling a comprehensive first-aid kit tailored for off-grid living, including both medical tools and herbal solutions for everyday health needs and emergencies.

Essential Medical Supplies

While natural remedies are invaluable in an off-grid environment, conventional first-aid supplies are still crucial for treating serious injuries. Below is a list of key items that should be included in any first-aid kit:

- **Bandages and Dressings**: For covering wounds and preventing infection.
    - Sterile gauze pads (various sizes).
    - Adhesive bandages (Band-Aids) in various sizes.
    - Elastic bandages for wrapping sprains.
    - Sterile adhesive strips for closing wounds (butterfly closures).
- **Wound Cleansing and Disinfection**:
    - **Alcohol wipes** or **hydrogen peroxide** for disinfecting wounds.
    - **Antiseptic ointment** (such as triple antibiotic ointment) for preventing infections.
    - **Sterile saline solution** for washing out wounds or eyes.
- **Instruments and Tools**:
    - Tweezers for removing splinters, ticks, or debris.
    - Small, sharp scissors for cutting bandages or clothing.
    - Safety pins for securing bandages or wraps.
    - Thermometer (preferably a digital or non-contact type).
- **Pain Relief and Medications**:
    - Over-the-counter pain relievers (such as **ibuprofen**, **acetaminophen**, or **aspirin**).
    - **Anti-inflammatory medications** (such as ibuprofen) for reducing pain and swelling.
    - **Antihistamines** (such as diphenhydramine) for allergic reactions.

Herbal Remedies for the Off-Grid First-Aid Kit

In addition to conventional supplies, herbal remedies provide a natural, sustainable alternative for treating common ailments when living off-grid. Below are some key herbs and remedies that should be included in your off-grid first-aid kit, as they can be easily grown or harvested from your garden:

- **Tea Tree Oil**:
    - A powerful natural antiseptic and antimicrobial oil that can be used to disinfect wounds, treat fungal infections, and soothe bug bites.

- **How to Use**: Apply a few drops to a clean cotton pad and dab directly onto cuts, scrapes, or insect bites to prevent infection.

- **Echinacea Tincture**:
  - Echinacea is a well-known immune booster and can help the body fight off colds, infections, and even speed up recovery from illness.
  - **How to Use**: At the first sign of illness, take a few drops of echinacea tincture in water or juice to strengthen the immune system.
- **Arnica Cream**:
  - Arnica is a natural anti-inflammatory that is excellent for bruises, sprains, and sore muscles.
  - **How to Use**: Apply arnica cream or salve directly to the affected area to reduce pain and swelling. It is not meant for use on open wounds.
- **Lavender Essential Oil**:
  - Lavender is a versatile herb with antiseptic, anti-inflammatory, and calming properties. It can be used to treat burns, cuts, and insect bites while also promoting relaxation and sleep.
  - **How to Use**: Apply diluted lavender oil (in a carrier oil) to minor burns or cuts. Add a few drops to a cloth or pillow for a calming effect to help with stress or anxiety.
- **Calendula Salve**:
  - Calendula is known for its skin-healing properties and can be used to treat wounds, rashes, and minor burns.
  - **How to Use**: Apply a thin layer of calendula salve to clean wounds, rashes, or areas of skin irritation to promote healing and reduce inflammation.
- **Plantain Leaves**:
  - Plantain is a common weed with powerful anti-inflammatory and astringent properties. It can be used to treat bug bites, stings, and wounds.
  - **How to Use**: Crush fresh plantain leaves to release the juice, then apply the crushed leaves directly to bites, stings, or small cuts as a poultice. Leave the poultice on for several hours, replacing it as needed.

. Remedies for Common Ailments

When living off-grid, you'll need to be prepared to manage common illnesses and ailments with limited resources. Below are some easy-to-make herbal remedies for addressing typical health problems such as colds, digestive issues, and stress.

- **Colds and Respiratory Infections**:
  - **Elderberry Syrup**: Elderberry is known for its ability to shorten the duration of colds and flu. You can make elderberry syrup by simmering dried elderberries with honey and water until thickened.
  - **Peppermint Tea**: Peppermint helps to relieve sinus congestion and soothe sore throats. Steep dried peppermint leaves in hot water and drink as needed for respiratory relief.
- **Digestive Discomfort**:
  - **Ginger Tea**: Ginger is an excellent remedy for nausea, indigestion, and bloating. Slice fresh ginger root and simmer in water for 10-15 minutes to make a soothing tea.
  - **Chamomile Tea**: Chamomile is gentle and effective for calming digestive upsets and reducing anxiety. Drink a cup of chamomile tea after meals or before bed to help with digestion and relaxation.

- **Pain and Inflammation**:
  - **Willow Bark**: Known as "nature's aspirin," willow bark contains salicin, which is similar to the active ingredient in aspirin. It can be used to relieve headaches, back pain, or joint pain. Simmer the bark in water to create a tea.
  - **Turmeric Paste**: Turmeric is a potent anti-inflammatory herb. You can make a paste by mixing turmeric powder with water and applying it directly to inflamed areas or wounds.
- **Stress and Sleep Problems**:
  - **Lavender Tea or Essential Oil**: Lavender is well-known for its calming properties and can help with anxiety and sleep disorders. Drink lavender tea before bed or place a few drops of lavender essential oil on your pillow.
  - **Valerian Root**: Valerian is a powerful herbal remedy for insomnia and anxiety. Make a tea from valerian root or take it in tincture form to promote relaxation and better sleep.

Emergency First-Aid Protocols

In more severe situations where medical care is not readily available, having a plan and understanding basic emergency protocols is crucial. Below are some first-aid guidelines for dealing with serious injuries and emergencies off-grid:

- **Wound Care and Infection Prevention**:
  - Clean all wounds immediately with saline solution or clean water.
  - Apply antiseptic (like tea tree oil or alcohol) and cover the wound with sterile gauze.
  - Monitor for signs of infection such as redness, swelling, or pus, and apply natural antibiotics like honey or garlic if needed.
- **Treating Burns**:
  - Cool the burn immediately with cool (not cold) water for 10-20 minutes.
  - Apply aloe vera gel or lavender oil to soothe and heal the skin.
  - For more severe burns, cover with sterile, non-stick bandages and seek professional care as soon as possible.
- **Dealing with Sprains or Fractures**:
  - Immobilize the injured area with a splint or sling if necessary.
  - Use arnica cream or comfrey salve to reduce swelling and pain.
  - For fractures, create a makeshift splint with available materials like sticks or boards, and seek medical attention when possible.

Storage and Maintenance of the Off-Grid First-Aid Kit

To ensure that your first-aid kit remains effective, it's important to regularly check and maintain its contents. Here are some tips for organizing and storing your off-grid first-aid kit:

- **Label Everything**: Clearly label all herbal remedies, tinctures, salves, and essential oils with their names, uses, and expiration dates.
- **Store Properly**: Keep the kit in a cool, dry place, away from direct sunlight, to prevent the degradation of herbs and tinctures. Essential oils should be stored in dark, glass bottles.
- **Restock Regularly**: Check your supplies every few months and restock any items that have been used or are approaching expiration. Fresh herbs should be replaced regularly to maintain potency.

20.2 Inspiring Others on the Path to Self-Sufficiency

Maintaining daily health and wellness in an off-grid environment is crucial for long-term survival and a high quality of life. Without access to modern healthcare, medications, or even regular grocery stores, you need to adopt a proactive approach to **nutrition**, **hygiene**, and **mental well-being**. In this section, we'll cover how to grow and sustain a **healthy diet**, maintain **personal hygiene** using natural ingredients, and implement strategies for **stress management and resilience**.

Growing a Nutritious Off-Grid Diet

In an off-grid setting, your diet will rely heavily on what you can **grow, forage, and preserve**. It's essential to ensure that your diet is balanced and provides all the nutrients you and your family need to stay healthy. Below are some key food groups you should focus on cultivating and preserving to maintain **daily health**.

- **Leafy Greens and Vegetables**:
  - **Spinach**, **kale**, **lettuce**, and **collard greens** are packed with essential vitamins like A, C, and K, as well as minerals like calcium and iron.
  - These plants are relatively easy to grow in raised beds or greenhouses, and they can be preserved by drying or freezing for long-term use.
  - **How to Use**: Incorporate leafy greens into soups, salads, and stews. You can also dehydrate them to make a nutrient-rich powder that can be added to meals during the winter months.
- **Root Vegetables**:
  - **Carrots**, **potatoes**, **sweet potatoes**, and **beets** are excellent sources of carbohydrates, fiber, and essential vitamins like beta-carotene and potassium.
  - Root vegetables are ideal for off-grid living because they store well in **root cellars** or cool, dark places, making them accessible throughout the year.
  - **How to Use**: Cook them in soups, roasts, or even mash them for an easy, filling side dish. Store them properly to ensure longevity during the colder months.
- **Legumes and Beans**:
  - **Beans**, **lentils**, and **peas** are vital sources of protein, especially when meat is not always available. They are also rich in fiber, making them essential for digestive health.
  - Beans can be grown in small spaces, dried, and stored for long-term use.
  - **How to Use**: Cook beans with herbs and vegetables for hearty stews, soups, or side dishes. They can also be sprouted for fresh greens if you lack access to fresh vegetables during winter.
- **Herbs and Medicinal Plants**:
  - Grow a variety of **culinary herbs** like **basil**, **oregano**, **thyme**, and **rosemary** to enhance the flavor of your meals and boost their medicinal value.
  - Medicinal herbs such as **echinacea**, **chamomile**, and **peppermint** can be grown to make teas, tinctures, and salves that support daily health and treat common ailments.

Personal Hygiene in an Off-Grid Setting

Good personal hygiene is essential to maintaining health in an off-grid environment, where the risk of infections and illnesses can be higher due to the lack of modern sanitation systems. Creating your own natural hygiene products from basic, sustainable ingredients will help you maintain cleanliness without relying on store-bought items.

- **Homemade Toothpaste**:
    - A simple, natural toothpaste can be made using just a few ingredients like **baking soda**, **coconut oil**, and **peppermint essential oil**.
    - **How to Make It**: Mix 2 tablespoons of baking soda with 2 tablespoons of coconut oil. Add 10-15 drops of peppermint oil for flavor and freshness. Store the mixture in a small glass jar and use it as you would regular toothpaste.
    - **Benefits**: Baking soda helps clean and whiten teeth, while coconut oil has antimicrobial properties that support oral health.
- **Natural Soap**:
    - Soap is essential for maintaining personal and household hygiene, and it can easily be made at home with natural ingredients like **lye**, **coconut oil**, **olive oil**, and **essential oils**.
    - **How to Make It**: Combine lye with water and allow it to cool. Mix the oils in a pot and heat them gently. Slowly add the lye to the oils, stirring until it thickens. Pour into molds and allow it to harden for 4-6 weeks before using.
    - **Benefits**: Homemade soap is free from harsh chemicals and fragrances, making it suitable for sensitive skin and environmentally friendly.
- **DIY Deodorant**:
    - Staying fresh in an off-grid setting doesn't require commercial deodorants filled with chemicals. You can make an effective natural deodorant using **baking soda**, **arrowroot powder**, and **coconut oil**.
    - **How to Make It**: Mix 1/4 cup of baking soda with 1/4 cup of arrowroot powder. Add 5-6 tablespoons of coconut oil and blend until smooth. Store in a small jar and apply as needed.
    - **Benefits**: This natural deodorant is effective in neutralizing odors and is gentle on the skin, making it a great alternative to store-bought options.
- **Rainwater Collection for Hygiene**:
    - If you are relying on rainwater or other off-grid water sources, ensure that water used for hygiene (such as for bathing and cleaning) is properly filtered or purified to avoid contamination.

Stress Management and Mental Wellness

Living off-grid can present numerous challenges, from food and water scarcity to the mental strain of isolation and unpredictability. Maintaining good **mental health** is just as important as physical health, especially in a survival scenario. Below are some strategies for **managing stress** and promoting mental wellness.

- **Meditation and Mindfulness**:
    - Practicing **meditation** or **mindfulness** techniques can help reduce stress and anxiety, improve focus, and build mental resilience. Set aside a few minutes each day to practice breathing exercises, guided meditation, or simply sit in silence and observe your surroundings.
    - **How to Practice Mindfulness**: Find a quiet spot, either inside or outside in nature. Focus on your breathing, taking slow, deep breaths. If your mind wanders, gently bring it back to your breath or the sounds of nature around you.
- **Physical Exercise**:

    - Regular **physical activity** is crucial for maintaining both physical and mental health. Incorporating daily exercise, even if it's just walking, gardening, or wood chopping, can significantly reduce stress levels and improve your mood.
    - **How to Stay Active**: Engage in outdoor activities that are functional, such as cutting wood, tending to your garden, or hiking. Not only will these activities improve your physical health, but they will also help reduce stress and improve your mental well-being.
- **Connecting with Nature**:
    - One of the greatest benefits of off-grid living is the close connection to nature. Spending time outdoors, whether gardening, hiking, or simply sitting in the sun, can help relieve stress and foster a sense of peace and well-being.
    - **How to Reconnect with Nature**: Make a daily habit of spending time outside, observing the changing seasons, wildlife, and plant growth. This practice can help reduce feelings of isolation and promote a deeper connection to your environment.
- **Maintaining a Routine**:
    - Establishing a daily routine can help maintain a sense of normalcy, structure, and control in an otherwise unpredictable environment. A simple routine that includes time for work, relaxation, and self-care can make a significant difference in mental well-being.
    - **How to Create a Routine**: Plan your day around essential tasks such as tending to livestock, gardening, and maintaining your home, but also schedule time for rest, hobbies, or creative pursuits to balance work with relaxation.

Herbal Remedies for Mental Wellness

In addition to lifestyle practices, you can use **herbal remedies** to support mental health and alleviate stress, anxiety, or insomnia. Below are some key herbs that you can grow and incorporate into your daily routine for mental wellness.

- **Chamomile**:
    - Chamomile is well-known for its calming effects and can be used to reduce stress, anxiety, and insomnia.
    - **How to Use It**: Drink chamomile tea in the evening to help unwind and promote restful sleep. You can also add dried chamomile flowers to a warm bath to relax the body and mind.
- **Lavender**:
    - Lavender is a powerful herb for promoting relaxation and reducing stress. Its soothing scent can help calm nerves and improve sleep quality.
    - **How to Use It**: Add a few drops of lavender essential oil to your pillow, diffuser, or bath. You can also make a lavender sachet and place it near your bed to enhance relaxation.
- **Valerian Root**:
    - Valerian is a potent herbal remedy for insomnia and anxiety. It helps promote relaxation and better sleep without the grogginess associated with some medications.
    - **How to Use It**: Make a tea or tincture from dried valerian root and consume it about 30 minutes before bed to improve sleep quality.

Conclusion: The Journey to Long-Term Off-Grid Self-Sufficiency

Living off the grid is more than just a survival strategy—it's a way of life that fosters independence, resilience, and a deep connection with the natural world. The projects and strategies outlined in this book are designed to help you build a sustainable, secure, and self-sufficient lifestyle, enabling you to thrive without reliance on modern infrastructure. By integrating these techniques, you can ensure that you and your family are prepared for anything, from temporary disruptions to long-term off-grid living.

Embrace the Challenges and Rewards of Off-Grid Living

Off-grid living is not without its challenges. You may face unpredictable weather, mechanical failures, or the isolation of living far from modern conveniences. However, by embracing these challenges, you also unlock the rewards of self-reliance, resourcefulness, and a simpler, more fulfilling lifestyle.

With each project you complete—whether it's setting up an energy system, growing your own food, or fortifying your home—you become more independent and resilient. These skills are not just for short-term survival; they are investments in your future, ensuring that you can meet your needs no matter what life throws your way.

A Lifelong Commitment to Sustainability

Achieving self-sufficiency is not a one-time project; it is an ongoing commitment. As you progress in your off-grid journey, continue to refine and improve your systems. Experiment with new techniques, scale up your operations, and explore ways to make your lifestyle more sustainable and comfortable. The ultimate goal is to create a system that not only supports your immediate needs but also continues to provide for you and your family in the long term.

Remember that self-sufficiency doesn't mean isolation. Engage with other off-grid communities, share your experiences, and learn from others. Building a network of like-minded individuals can offer support, encouragement, and opportunities for trade or barter.

The Path Forward: Continuous Learning and Adaptation

The path to off-grid living is never truly finished. There will always be new challenges to overcome and opportunities to enhance your systems. As you adapt to changing conditions—whether they be environmental, economic, or personal—you'll continue to grow and improve your homestead. Your journey toward self-reliance is an evolving one, and with the knowledge and skills gained, you'll be well-prepared for whatever comes next.

Sharing Your Experiences

As you build your off-grid life, consider sharing your knowledge with others. Whether it's through local workshops, online forums, or simply helping a neighbor, your experiences can inspire others to start their own journey toward self-sufficiency. By contributing to the broader community, you not only help others become more resilient but also strengthen your own support network.

Final Thoughts

Living off the grid is a bold and rewarding choice. It empowers you to take control of your resources, your environment, and your future. While the road may be difficult at times, the sense of accomplishment, independence, and security you gain is unmatched.

As you continue on this path, remember that every challenge is an opportunity to grow, and every project completed brings you one step closer to true self-sufficiency. The tools, strategies, and mindset outlined in this book are designed to support you in this journey, helping you build a resilient, sustainable, and fulfilling off-grid life.

Here's to your success in creating a life of independence, sustainability, and peace—living off the grid, on your own terms.

Bonus: Step-by-Step DIY Projects for Off-Grid Independence

Discover a series of practical, detailed projects to enhance your off-grid self-sufficiency. Each bonus will guide you step-by-step through the creation of essential tools for independent living:

Made in the USA
Thornton, CO
03/24/25 10:26:12

40242dc3-9115-42d9-aab0-ce57e90826c6R01